To:

Steffi, Angela, Gianni, Piero, Nona, Chicca
and all the people who supported me through the years of writing.

For other titles in the Series see p. 245

INFORMATION TECHNOLOGY & LAW SERIES (17)

TRUSTMARKS IN E-COMMERCE

The Value of Web Seals and the Liability of their Providers

Paolo Balboni

T·M·C· ASSER PRESS
The Hague

The *Information Technology & Law Series* is published
by T·M·C·ASSER PRESS
P.O. Box 16163, 2500 BD The Hague, The Netherlands
<www.asserpress.nl>

T·M·C·ASSER PRESS English language books are distributed exclusively by:

Cambridge University Press, The Edinburgh Building, Shaftesbury Road,
Cambridge CB2 2RU, UK,
or
for customers in the USA, Canada and Mexico:
Cambridge University Press, 100 Brook Hill Drive, West Nyack, NY 10994-2133, USA

<www.cambridge.org>

The *Information Technology & Law Series* is an initiative of IT*e*R, the National Programme for Information Technology and Law, which is a research programme set up by the Dutch government and the Netherlands Organisation for Scientific Research (NWO) in The Hague. Since 1995 IT*e*R has published all of its research results in its own book series. In 2002 IT*e*R launched the present internationally orientated and English language *Information Technology & Law Series*. This series deals with the implications of information technology for legal systems and institutions. It is not restricted to publishing IT*e*R's research results. Hence, authors are invited and encouraged to submit their manuscripts for inclusion. Manuscripts and related correspondence can be sent to the Series' Editorial Office, which will also gladly provide more information concerning editorial standards and procedures.

Editorial Office
eLaw@Leiden, Centre for Law in the Information Society
Leiden University
P.O. Box 9520
2300 RA Leiden, The Netherlands
Tel. +31(0)715277846
E-mail: <ital@law.leidenuniv.nl>
Web site: <www.nwo.nl/iter>

Single copies or Standing Order
The books in the *Information Technology & Law Series* can either be purchased as single copies or through a standing order. For ordering information see the information on top of this page or visit the publisher's web site at <www.asserpress.nl/cata/itlaw17/fra.htm>.

ISBN 978-90-6704-296-3
ISSN 1570-2782

Cover and lay-out: Oasis Productions, Nieuwerkerk a/d IJssel, The Netherlands

FOREWORD

The present book touches upon crucial and extremely timely issues of ICT development and application: privacy, security, and e-consumers' trust. It does so by dealing with the challenges posed by the instrument of trustmarks. Trustmarks, in a nutshell, are electronic labels or visual representations indicating that an e-merchant has demonstrated its conformity to standards regarding security, privacy, and business practice. From a business perspective trustmarks can be seen as a way to convey e-consumers' trust. In fact, e-merchants hope that by displaying trustmarks on their websites, e-consumers will less likely question the integrity of such e-merchants in relation to security, privacy, and business practice and thus be more likely to divulge their personal data and transact concerning a service or product. From a regulatory perspective, trustmarks can also be seen as a way of self-regulating or co-regulating the e-market on privacy and security issues.

Indeed this subject is what makes the book unique, being the first comprehensive study on trustmarks. On the one hand, the author focuses on legal aspects, paying special attention to the possible liability of Trustmark Organisations, 'TMOs' (i.e., the entities that issue trustmarks). On the other hand, by a multidisciplinary approach to the matter, Balboni describes the pros and cons of trustmarks for e-economy, e-society, and e-policy.

The book starts off by explaining what trustmarks are, how TMOs issue trustmarks, and to what extent this instrument may contribute to enhancing e-consumers' trust. The input for the observations and conclusions is provided through an analysis of the practices of four TMOs based in the US and five TMOs based in Europe. This analysis is combined with a thorough investigation of the relevant international literature on the issue.

The author then elaborates – by carrying out a critical comparative research into the American, English, German, and French legal systems – on the legal value of trustmarks and TMOs' liabilities towards e-consumers who have suffered a loss after relying on inaccurate trustmarks.

The book ends with recommendations for an innovative legal framework for trustmarks which aims at enhancing e-consumers' trust in e-commerce. Of particular interest is the manner in which the author applies the ethical theory of 'Warranted Trust' to the relationship between e-consumers and e-merchants in order to fine-tune his proposals.

The volume is the result of a four-year multidisciplinary comparative research based at Tilburg University (the Netherlands). During this time the author framed his manuscript by constantly confronting his findings, views, and theories not only with his colleagues from the Department of Private Law at the Tilburg Institute for Law, Technology, and Society (TILT), and his fellows at the University of Bologna (Italy) but also by presenting papers on the topic at major international ICT and law conferences.

This book is a must for those working in regulatory bodies involved in ICT, TMOs, e-merchants, e-consumers, e-consumers' associations, ICT lawyers, academics and, more generally, for anyone who wants to have a complete and accurate insight into the role trustmarks may play in enhancing e-commerce.

Tilburg, the Netherlands, February 2009 Corien PRINS*

---

* Professor of law and informatisation at the Tilburg Institute for Law, Technology, and Society (TILT), Tilburg University, and Council Member of the Scientific Council for Government Policy (WRR), the Netherlands

TABLE OF CONTENTS

ABBREVIATIONS

A.2d	Atlantic Reporter, Second Series
A.C.	Law Reports, Appeal Cases
AECE	Asociación Española de Comercio Electrónico
AICPA	American Institute of Certified Accountants
AUTOCONTROL	Asociación para la Autorregulación de la Comunicación Comercial
BBB	Better Business Bureau
BGB	Bürgerliches Gesetzbuch
BGBl	Bundesgesetzblatt
BGH	Bundesgerichtshof
BGHZ	Bundesgerichtshofs in Zivilsachen
BEUC	Bureau Européen des Unions de Consommateurs
BSE	Bovine Spongiform Encephalopathy
Bull. C.N.C.C.	Bulletin du Conseil National des Commissaires aux Comptes
Bull. Civ.	Bulletin des arrêts de la Chambre Civile de la Cour de Cassation
Bull. Joly	Bulletin Joly (mensuel d'information des sociétés)
Cal.	California Reports (Supreme Court)
Cal. Rptr.	West's California Reporter
Cass	Cour de cassation
Cass. Ass. plén.	Arrêt de l'assemblée plénière de la Cour de Cassation
Cass.com.	Cour de cassation (Chambre commerciale et financière)
CICA	Canadian Institute of Chartered Accountants
Circ.	Circuit Court of Appeals (federal)
Com.	Arrêt de la chambre commerciale de la cour de cassation
CSP	Certification Service Provider
D. aff.	Recueil Dalloz, édition « affaires » (1995-1999)
D. R.I.	Federal District Court Rode Island
DIT	Revue droit informatique et télécoms
DNS	Domain Name System
EC	European Community
ENISA	European Agency on Information Security
EU	European Union
ETR	European Trustmark Requirements

F.2d	Federal Reporter, Second Series
F. Supp.	Federal Reporter Supplement
FAQ	Frequent Asked Question
FMD	Foot and Mouth Disease
FTC	Federal Trade Commission
GAO	General Accounting Office
Gaz. Pal.	Gazette du palais
HL	House of Lords
HGB	Handelsgesetzbuch
Id.	Idem
Ind. Ct. App.	Indiana Court of Appeal
INPI	Institut national de la propriété industrielle
IR	Informations rapides du recueil Dalloz
ISO	International Organisation for Standardisation
IT	Information Technology
JCP	Jurisclasseur périodique
JO	Journal officiel
JuS	Juristische Schulung
JZ	Juristen-Zeitung
K.B.	Law Reports, King's Bench
L.J.	Lord Justice
La.	Louisiana Reports
Lloyd's L. Rep. P.N.	Loyd's Law Reports Professional Negligence
Loyid's Rep.	Lloyd's Law Reports
LQR	Law Quarterly Review
Md.	Maryland Reports
N.C.	North Carolina Reports
N.E.	North Eastern Reporter
N.E.2d	North Eastern Reporter, Second Series
N.H.	New Hampshire Reports
N.J.	New Jersey Reports
N.J. Stat. Ann.	New Jersey Statutes Annotated
NJW	Neue Juristische Wochenschrift
N.W.2d	North Western Reporter, Second Series
N.Y.	New York Reports
N.Y. App. Div.	New York Supreme Court, Appellate Division
N.Y.S.	New York Supplement Reporter
N.Y.S.2d	New York Supplement Reporter, Second Series

Ohio	Ohio Reports
OJ	Official Journal
OLAS	Office Luxembourgeois d'Accréditation et de Surveillance
OLG	Oberlandesgericht
P.2d	Pacific Reporter, Second Series
Pa.	Pennsylvania State Reports
PNLR	Professional Negligence and Liability Reports
RD bancaire bourse	Revue de droit bancaire et de la bourse
RTD civ.	Revue trimestrielle de droit civil
RTD com.	Revue trimestrielle de droit commercial
Rev. Soc.	Revue des sociétés Dalloz
Riv. Cr. Dir. Priv.	Rivista critica del diritto privato
S.E.2d	South Eastern Reporter, Second Series
Sag	Erstes Gesetz zur Änderung des Signaturgesetzes
sgA	Stillschweigend geschlossener Auskunftsvertrag
S.W.2d	Southern Western Reporter, Second Series
S&P	Standard & Poor's
SNCH	Société Nationale de Certification et d'Homologation
So. 2d	Southern reporter, Second Series
StGB	Strafgesetzbuch
Sup. Ct.	Supreme Court
Tenn.	Tennessee Reports
Tex. App.	Texas Appeals Reports
TMO	Trustmark Organisation
Trib. Gr. Inst.	Tribunal de grande instance
TRUSTe	Trusted Universal Standards in Electronic Commerce
TTP	Trusted Third Party
UNICE	Union of Industrial and Employers Confederations in Europe
U.S.	United States Supreme Court Reports
US	United States
VmSzD	Vertrag mit Schutzwirkung zugunsten Dritter
VR	Victorian Reports
VzD	Vertrag zugunsten Dritter
W. Va.	West Virginia Reports
Wash. App.	Washington Appellate Reports
WDPR	World Data Protection Report
Wis.	Wisconsin Reports
W.L.R.	Weekly Law Reports
WM	Zeitschrift für Wirtschafts- und Bankrecht

ZfBR	Zeitschrift für deutsches und internationales Bau- und Vergaberecht

Chapter 1
INTRODUCTION*

1. Background

A few years after the commercialisation of the Internet, it became clear to businesses that the challenge for the coming years would be to win the trust of the consumers. First, consumers need to trust the Internet in general as a new means of communication. Second, consumers need to trust e-commerce as a new way of transferring ownership or rights to use goods or services by making transactions through a computer-mediated network. Third, consumers need to trust the business practices of a specific company which sells goods or provides services online in order to choose and thus do business with that specific company instead of preferring to deal with another one out of an almost unlimited number of options available online.[1]

Speaking about trust is comparable to speaking about love: one can talk for hours without reaching any definite conclusion. Trust is a feeling, so it is really difficult to measure and to identify all the factors that affect it.[2] However, as with love, there are some basic rules which, if followed, can at least increase the chances to enhance trust. These rules vary and transform together with the development of society. Nowadays, security and privacy seem to be at the top of the list of consumers' concerns in online transac-

---

* All the websites quoted in this book were last visited on 11 September 2008.

[1] Improving the security of the electronic transactions, in particular in e-commerce, and increasing consumers' trust and confidence in them have a high priority on the political agenda of the European Union. See 'i2010 – A European Information Society for growth and employment'. Available at <http://ec.europa.eu/information_society/eeurope/i2010/index_en.htm>.

[2] See Cross F. (2005), p. 1461. 'While we have a generalized understanding of the concept of trust, it is not readily amenable to clear definition. (...) Research has at least implicitly accepted a definition of trust as a belief, attitude, or explanation concerning the likelihood that the actions or outcomes of another individual, group or organisation will be acceptable (...) or will serve the actor's interests.'

P. Dalboni, Trustmarks in E-Commerce

tions.[3] If there is concern, trust is unlikely to develop. Therefore, if online companies want to gain consumers' trust, they have to start by presenting the consumers with trustworthy security measures and a trustworthy privacy policy.

Starting from the late 1990s in the US and a few years later also in Europe, a potential solution to the need for trust of e-consumers and e-merchants has been offered by a sort of quality certification system provided by Trustmark Organisations (henceforth: TMOs). 'Certification is a procedure by which a third party gives a written assurance that a product, process or service conforms to specific characteristics.'[4] TMOs – also defined as Trusted Third Parties (henceforth: TTPs),[5] or parties that can be trusted and relied on – are, in a nutshell, organisations which present themselves as independent parties (third parties) and provide trustmarks (a label or visual representation indicating that a product, process or service conforms to specific quality characteristics) to online merchants (henceforth: e-merchants).[6]

Security, privacy, and more generally business practice[7] are the fields in which trustmarks are very popular. In fact, a TMO will issue a trustmark to an e-merchant that has demonstrated its conformity to the policy standards of the TMO regarding security, privacy, and business practice. The e-merchant hopes that by displaying the trustmark on its website, e-consumers will less likely question the integrity of that e-merchant in relation to security, privacy, and business practice. Thus, the consumer will be more likely to divulge his personal data to and transact with that e-merchant, i.e., to conduct online business.

---

[3] See Organisation for Economic Co-operation and Development – Working Party on Indicators for the Information Society (2005), *Scoping Study for the Measurement of Trust in the Online Environment*, DSTI/ICCP/IIS(2005)1/FINAL. Available at <www.oecd.org/dataoecd/26/15/35792806.pdf>; Organisation for Economic Co-operation and Development – Working Party on Indicators for the Information Society (2008), p. 8. Available at: <http://www.oecd.org/dataoecd/47/18/40009578.pdf>; *Eurobarometer* (2004), p. 5; *Eurobarometer* (2003), p. 44.

[4] Rae, A. et al. (1995), p. 2. For more information on the importance of certification for e-commerce and the related legal issues, see Brun, M.B. (2001).

[5] Froomkin, M. (1996), p. 49. See also Duthler, A.W. (1998).

[6] See more on TMOs and their practice in Chapter 2 *infra*.

[7] Business practice is a term which has a very broad meaning and embraces various e-merchant practices, e.g., security, privacy, marketing, information.

In summary, TMOs are independent organisations which try to promote online trust by offering a system of certification.

So far so good. Now let us take a close look at the trustmark system and how it works in reality. Such system is generally based on co-operation between TMOs and e-merchants that ask for the trustmark. More precisely, the e-merchant should check its security, privacy or business practice and then provide a self-assessment with the understanding and expectation that the TMO may rely on the statement contained therein for the purposes of determining whether the e-merchant's practice complies with the TMO standards.[8] However, when the co-operation between TMOs and e-merchants does not work, the trustmark system shows its weaknesses. In fact, TMOs have already proven to be not as trustworthy as they seem to be. There have been cases in which e-consumers' data were kept, shared, or sold by e-merchants without obtaining the data subjects' prior consent and without the knowledge of the TMO which issued the trustmarks to such e-merchants. The trustmark was on the e-merchants' website at the time the violations occurred and remained there after the wrongful act was discovered.[9] Therefore, it may become dangerous to trust the trustmarks. The chances of undetected e-merchant practices that are not in compliance with the TMOs' programmes are high, and malpractice can cause damage to e-consumers. As it will be further explained in Section 2, the damage can range from violation of e-consumer's privacy and data protection right[10] to pure economic loss.

The pictured scenario raises the following question, which is of evident practical relevance:

What can e-consumers do in order to recover the damage suffered from relying on a trustmark?

In order to elaborate the answer to such question, the legal relationships between the players of the certification system offered by TMOs need to be

[8] See more extensively on TMO practice in Chapter 3, Section 5 *infra*.

[9] See Chapter 2, Section 6 *infra*.

[10] Generally recognised in all Europe through the implementation of the Directive 95/46/EC of the European Parliament and of the Council of 24 October 1995 on the protection of individuals with regard to the processing of personal data and on the free movement of such data, *OJ* L 281, 23.11.1995, p. 31.

spelled out first. Accordingly, there is a contractual relationship between TMOs and their clients (e-merchants). A contract may usually occur also between the certified e-companies and their clients (e-consumers). However, there seems to be a tortious relationship between TMOs and e-consumers that rely on the certificates, although a contractual relationship cannot be excluded *a priori* (see Table 1).

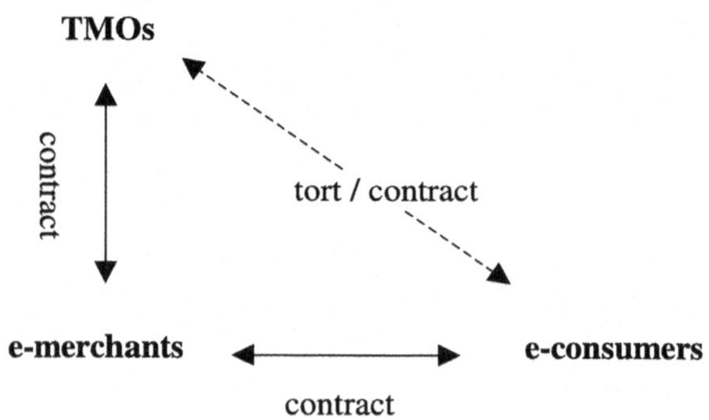

Table 1. Legal relationships

In order to recover the damage, e-consumers may sue the e-merchant for breach of a contractual obligation or, in tort, for wilful act or breach of their duty of care (i.e., negligence liability). However, it will not always be that easy. De-materialisation of personal identities and internationalisation,[11] two inner features of online communications, can make it very difficult for an e-consumer to actually identify and localise an e-merchant. Practically anybody, under false identity, and from any place in the world, can register a domain name, set up a commercial website overnight, and start selling goods or offering services worldwide through the Internet. Moreover, as it is very easy and quick for an e-merchant to set up a website, it is also easy and quick to shut it down and disappear without leaving any trace. This will decrease the chance for an e-consumer to enforce his rights against malicious or negligent e-merchants. It has to be pointed out that the players of the pictured scenario will most likely be small e-merchants, as opposed to big ones (i.e., companies which regularly offer their products and services

[11] See Chapter 2, Section 3 *infra*.

online). Trustmarks, as a means to gain credibility, are much more appealing to small e-merchants than to big ones, as the latter very often already have a consolidated reputation and clientele.

Given the described setting, theoretically, it would be easier for e-consumers to seek redress directly from the TMO who issued the trustmark to the e-merchant than from the e-merchant itself. First, the TMO should be easier to localise. Its contact details should be clearly stated in the website. Second, it should have more money to satisfy the e-consumers' request for compensation – as TMOs should generally be better capitalised than the small e-merchants that they certify. E-consumers may sue the TMO for the provision of inaccurate information on the e-merchant's practices. The TMO, on its side, could then seek redress from the e-merchant whose identity and location should be known to the TMO because of the certification procedure they had been through.

The relevant question here is whether the TMO may be held liable by aggrieved e-consumers.

None of the hard[12] and soft laws[13] applicable to TMOs sets forth specific rules on TMO liability. Apparently, governments have so far opted for

---

[12] E.g., Directive 2005/29/EC of the European Parliament and of the Council of 11 May 2005 concerning unfair business-to-consumer commercial practice in the internal market and amending Council Directive 84/450/EEC, Directives 97/7/EC, 98/27/C and 2002/65/EC of the European Parliament and of the Council and Regulation (EC) No 2006/2004 of the European Parliament and of the Council ('Unfair commercial Practices Directive'), *OJ* L 149, 11.6.2005, p. 22; Directive 2000/31/EC of the European Parliament and of the Council of 8 June 2000 on certain legal aspects of information society services, in the Internal Market (Directive on electronic commerce), *OJ* L 178, 17.7.2000, p. 1; Directive 97/7/EC of the European Parliament and of the Council of 20 May 1997 on the protection of consumers in respect of distance contracts, *OJ* L 144 , 4.6.1997, p. 19; Directive 95/46/EC of the European Parliament and of the Council of 24 October 1995 on the protection of individuals with regard to the processing of personal data and on the free movement of such data, *OJ* L 281, 23.11.1995, p. 31.

[13] E.g., European Trustmark Requirements (ETR). Available at <http://www.quatro-project.org/files/file/unice-beuc/eConfidence.pdf>. The UNICE–BEUC e-Confidence project establishes European trustmark requirements (ETR) aiming to provide a high standard of consumer protection in electronic commerce and encourage the sale of goods and services on the Internet. The ETR offer a basis for good online practice. They do not seek to override or replace any mandatory provisions at European level. They are supplementary to legal obligations and do not affect consumers' statutory rights. BEUC, the European Consumers' Organisation, is the Brussels-based federation of independent national consumer organisations from all the Member States of the EU and from other European countries. See <http://

a non-intervention policy in the trustmark sector. Moreover, there is (almost) no literature on the liability of TMOs. More generally, '[l]iterature has discussed issues such as privacy law, intellectual property regimes, licenses for selling drugs and medicines and finally, regulation on commercial communications and advertisements. None of these regulations, however, offer direct protection against damages caused by reliance on information.'[14]

The absence of specific liability rules for online service providers, such as TMOs, does not come as a surprise. The Internet has created many new services and the legislator has not regulated all of them, sometimes to leave room for self-regulation, other times because it is difficult or simply because regulation needs time and money. However, this does not mean that there is no liability for 'unregulated' online service providers and, in the present case, for TMOs. If there is no specific regulation, the general principles of tort and contract law will be applicable. In fact, the main instrument for protection against damages caused by reliance on inaccurate information is civil liability.[15] Moreover, the principle of analogy can be applied in order to use the rules set forth for offline and online professionals which are comparable to TMOs.[16] However, TMOs usually tend to limit or even exclude their liability towards both e-merchants and third parties through contractual provisions.[17]

2. Research questions

Given the scenario described above, it is possible to sketch what a typical TMO third-party liability case may look like (henceforth: the typical TMO third-party liability case).

An e-consumer, relying on a security-, privacy- or business practice-trustmark placed on an e-merchant website, decides to interact with such e-

ec.europa.eu/consumers/cons_int/e-commerce/e-conf_working_doc.pdf>; Principles for e-Commerce Codes of Conduct – Second Draft (The E-Confidence Initiative Working Documents); Global Business Dialogue on e-Commerce Recommendations.

[14] Schellekens, M. & Prins, C. (2006), p. 51.

[15] See Id.

[16] See Subsection 3.1 *infra*.

[17] See Chapter 3, Subsection 5.5 *infra*.

merchant. The e-consumer purchases, for example, goods from the e-merchant website or subscribes to an e-merchant's online service. To complete the transaction, the e-consumer provides the personal data the e-merchant asks for (e.g., his name, surname, date of birth, phone number, address, e-mail address, interests, purchase preferences), together with the relevant payment details (e.g., credit card number, bank details). In fact:

a) the e-consumer does not receive the good or the service he has paid for;
b) without the e-consumer's prior consent, his personal data are processed by the e-merchant for purposes other than the fulfilment of the relevant contractual obligations (e.g., used for profiling- and marketing-related purposes, shared or sold to third parties) and eventually the e-consumer starts to receive unsolicited marketing e-mails and phone calls by the e-merchant and/or by third parties; and
c) e-consumer's payment details are used directly by the e-merchant to defraud the e-consumer; shared or sold by the e-merchant to third parties who ultimately defraud the e-consumer; or stolen during the transaction or from the e-merchant's client database by a cunning third party who takes advantage of the poor security of the e-merchant's IT infrastructure.

In any or all of the circumstances indicated in items a, b, and c above, he will suffer damage. The damage can range from violation of the e-consumer's privacy and data protection right[18] to pure economic loss.

For the reason mentioned in the previous section, it may be difficult to recover the damage from the e-merchant. Hence, it is relevant to check whether the e-consumer may recover the damage directly from the TMO that issued the trustmark to the e-merchant and on what legal basis.

From the 'typical TMO third-party liability case' originates the following legal question:

a) *What is TMO third-party liability in Europe?* More precisely: *Are TMOs liable towards e-consumers who detrimentally rely on inaccu-*

[18] Generally recognised in all of Europe through the implementation of the Directive 95/46/EC of the European Parliament and of the Council of 24 October 1995 on the protection of individuals with regard to the processing of personal data and on the free movement of such data, *OJ* L 281, 23.11.1995, p. 31.

rate trustmarks and suffer loss? From a different perspective: *Do e-consumers have a cause of action to recover the damages they incurred by relying on inaccurate trustmarks directly from TMOs? If they do, what will be the legal basis?*

Moreover, given the TMO's role in e-commerce and the related trustmark potentialities, it will be interesting to evaluate the present liability system by answering the question:

b) *Is the present TMO third-party liability system in Europe adequate?*[19]

Eventually, if the TMO third-party liability system in Europe proves to be inadequate, it will be logical to bring the matter one step further and to provide an answer to the question:

c) *What will an adequate TMO third-party liability system be?*

These are the three fundamental questions that the present book aims to answer.

3. Methodology

The book starts off by reviewing the major issues that e-consumers face in e-commerce and demonstrating how trustmarks can – theoretically – serve e-consumers' primary need for reliable information and, at the same time, offer benefits to e-merchants and governments. However, also with the support of a case study, it is provisionally concluded that, both in US and in Europe, the present TMO practice is generally not trustworthy. Instead of contributing to enhancing e-consumer's, trust in e-commerce, this may actually represent a risk to e-consumers, e-merchants and governments and eventually undermine the growth of e-commerce (Chapter 2).

---

[19] The adequacy of the third-party liability system will be assessed considering TMO practice, its impact on e-consumers, e-merchants, governments, as well as on the same TMOs, and the deterring role that liability should have – it being one of the first legal guards against undesirable societal behaviour. See Section 3 and Chapter 9, Section 4 *infra*.

This provisional conclusion is confirmed by an in-depth analysis of four TMOs based in the US and five TMOs based in Europe. Although this book is about European TMO practice, American TMOs are dealt with for mainly three reasons. First, because they have influenced European TMO practice very significantly. Second, because some of them offer their services also in Europe. Third, because the whole trustmark phenomenon was started by the four selected TMOs in the US. Hence, the analysis of their practice is needed in order to have a better and broader understanding of the matter. The selection criteria differ between American and European TMOs. The US market is dominated by the four selected TMOs. Therefore, they are a representative sample of the US practice. The European market, however, consists of many small players. None of them has reached critical mass. Moreover, it is usually not so easy to find much information on their practice. The five TMOs are thus chosen because they made enough information available on their websites to carry out an accurate analysis. As to the analysis criteria of the selected TMOs' practice, first, the certification process is defined; second, the necessary conditions for a trustworthy certification practice are isolated; and third, the practice of the nine benchmarked TMOs is evaluated against the necessary conditions for a trustworthy certification practice (Chapter 3).

3.1 Comparative legal analysis

Having explained what TMOs are, the service they provide, and the current state of the art of their practice, the comparative analysis of their third-party liability will take place (Chapters 4, 5, 6, and 7).

3.1.1 *Choice of the legal systems to compare*

The fundamental questions that this book aims to answer are related to Europe.[20] Therefore, following a classical method of private comparative law, the English, the German, and the French legal systems will be analysed to obtain insights on the most representative legal systems in Europe. In fact, according to the theory of 'legal families', which try to bring back the vast number of legal systems into few large groups (the 'legal families'),

[20] See Section 2 *supra*.

the Romanistic legal family is well represented by the French legal system, the Germanic by the German one, and the Anglo-American by the English legal system.[21] However, because the trustmark phenomenon started in the US and there it is more expanded than that in Europe, as a matter of completeness, the American legal system will also be analysed. In this way, the reader will have a better understanding of the legal issues related to trustmarks. Furthermore, as already pointed out, there are a number of well-established American TMOs which also operate in Europe, bringing along their own terms and conditions for the services they offer in the old continent and influencing European TMOs' legal practice.[22] Last but not least, since the second and the third research questions are evaluative in nature,[23] knowledge of a non-European system will allow for more powerful conclusions and serve as an additional source of inspiration.

3.1.2 *Choice of TMO offline equivalents*

Given that in the analysed legal systems there are no specific rules on TMO liability, following the principle of 'what applies offline should also apply online',[24] which has been generally used to fill the legislative gaps on online matters, offline professionals comparable to TMOs are identified in surveyors, auditors, and accountants. More precisely, following the principle of functionality,[25] trustmarks are seen as information on somebody or something to be relied upon by others. Accordingly, TMOs are seen as professionals who provide information on their clients, or their clients' practice, to be relied upon by third parties (e.g., e-consumers). Surveyors, auditors and accountants may also be seen as professionals who provide information on somebody or something to be relied upon by others. (The comparison amongst TMOs, surveyors, and auditors/accountants is further specified in Table 2.)

---

[21] Zweigert, K. & Kötz, H. (1998), pp. 63et seq.

[22] See Chapter 3, Sections 4 and 5 *infra*.

[23] See Section 2 *supra*.

[24] See, e.g., Schellekens, M. (2006), pp. 51-76; Koops, B.J. et al. (2000).

[25] The principle of functionality is the basic methodological principle of all comparative law analysis. 'Incomparables cannot usefully be compared, and in law the only things which are comparable are those which fulfil the same function.' Zweigert, K. & Kötz, H. (1998), p. 34. For further information on the principle of functionality, see also Id., pp. 34 s. and 44; Ralf, M. (2006), pp. 339-382.

	TMOs	*Surveyors*	*Auditors/Accounts*
Independent and trustworthy info	Trustmarks	Valuations	Reports
On the quality of goods or practice	Security, privacy, and business practices	Real estate	Accounts
To be relied upon by third parties	E-consumers	Purchasers	Shareholders/investors

Table 2. Table of comparison

The third-party liability regime of surveryors, auditors, and accountants will be studied through the analysis of the relevant provisions and case law. Moreover, its applicability by analogy to TMOs and the related effects on TMO third-party liability will be investigated. However, it has to be pointed out that the analysis of significant cases which do not concern surveyors, auditors, and accountants but are nevertheless related to third-party liability for the provision of information will sometimes be allowed in order to have a broader and more complete view of the issue at stake.

One could ask why surveyors, auditors, and accountants are preferred to certifiers as offline professionals comparable to TMOs. There are two main reasons for this choice. First, certifiers are actually a species of the broader genus of the auditors, as in practice certifiers carry out auditing activities.[26] Therefore, the choice has been to study the issue from a broader perspective.[27] Second, instead of focusing on offline certifiers, the book deals with online certifier third-party liability, which is more relevant to the present analysis. Given the implementation in Europe of Article 6 of the Directive of the European Parliament and of the Council of 13 December 1999 on a Community Framework for Electronic Signatures (99/93/EC),[28] which sets forth a third-party liability clause for Certification Service Providers (henceforth: CSPs) – professionals who, for the services they provide, are compa-

[26] See Chapter 3, Subsection 2.2 *infra.*

[27] Moreover, it has to be noticed that certifier third-party liability has been generally denied in Europe. Only contracting party can enforce damage payment related to losses suffered relying on the certificates. See e.g., Jahn, G. et al. (2005), p. 64.

[28] Directive of the European Parliament and of the Council of 13 December 1999 on a Community Framework for Electronic Signatures (99/93/EC), *OJ* L 13, 19 January 2000, p. 12.

rable by analogy to TMOs – the effects of the application by analogy of such provisions to TMOs are dealt with in the analysis of the European legal systems.

3.1.3 *Possible application by analogy of Certification Service Providers liability rules to TMOs*

According to Article 2 (11) of the Electronic Signatures Directive, a CSP is 'an entity or a legal or natural person who issues certificates or provides other services related to electronic signatures.' CSPs are TTPs[29] that offer mainly three services: authentication, time-stamping, and certification management. First, the authentication service consists of linking the signatory ('a person who holds signature creation device and acts either on his own behalf or on behalf of the person he represents'[30]) to his digital signature. In practice, the exchange of digital certificates (provided by CSPs), through an automatic 'digital handshake' between computers, provides assurance that the parties are who they say they are and helps assess whether the service provided and the goods or services delivered are genuine. Second, time stamping means to create a notation that indicates the correct date and time of an action. Through the time-stamping service, CSPs guarantee the time at which an electronic document was digitally signed. Third, CSPs offer certification management services. In fact, for a number for reasons, the certificate may lose its trustworthiness or become unreliable[31] and the CSP (at the signatory's request or even without the signatory's consent, depending on the circumstances) may suspend (temporarily interrupt the operational period) or revoke (permanently invalidate) the certificate. Immediately upon suspending or revoking a certificate, the CSP is expected to publish a notice of the revocation or suspension or notify persons who have enquired or are known to have received a digital signature verifiable by reference to the unreliable certificate.

---

[29] On Trusted Third Parties, see Section 1. For more information on CSPs as Trusted Third Parties, see Froomkin, A.M. (1996), pp. 49 et seq.

[30] Article 2 (3) of the Electronic Signatures Directive.

[31] E.g., in a situation where the signatory misrepresents its identity to the CSP. In other circumstances, a certificate may be reliable enough when issued, but sometime thereafter, it may become unreliable (e.g., the private key is compromised through loss of control of it by the signatory).

Courts could compare TMOs with CSPs because they are two species of the broader genus of TTPs. Moreover, CSPs fit the definition given in Section 3 of the TMOs' equivalent. CSPs are professionals that provide information on somebody (i.e., the identity of the signatory) or something (i.e., the time at which an electronic document was digitally signed and the validity of the certificate) to be relied upon by others (third parties who are recipients of digitally signed messages).

The third-party liability of CSPs has been set out in Article 6[32] of the Electronic Signatures Directive.[33] The implementation of this provision in the English, German, and French legal systems and its potential impact on TMO third-party liability will thus be analysed.

---

[32] Article 6. Liability. 1. 'As a minimum, Member States shall ensure that by issuing a certificate as a qualified certificate to the public or by guaranteeing such a certificate to the public a certification service provider is liable for damage caused to any entity or legal or natural person who reasonably relies on that certificate: (a) as regards the accuracy at the time of issuance of all information contained in the qualified certificate and as regards the fact that the certificate contains all the details prescribed for a qualified certificate; (b) for assurance that at the time of the issuance of the certificate, the signatory identified in the qualified certificate held the signature-creation data corresponding to the signature-verification data given or identified in the certificate; (c) for assurance that the signature-creation data and the signature-verification data can be used in a complementary manner in cases where the certification service provider generates them both; unless the certification service provider proves that he has not acted negligently. 2. As a minimum, Member States shall ensure that a certification service provider who has issued a certificate as a qualified certificate to the public is liable for damage to any entity or legal or natural person who reasonably relies on the certificate for failure to register revocation of the certificate unless the certification service provider proves that he has not acted negligently. 3. Member States shall ensure that a certification service provider may indicate in a qualified certificate limitations on the use of that certificate, provided that the limitations are recognisable to third parties. The certification service provider shall not be liable for damage arising from use of a qualified certificate which exceeds the limitations placed on it. 4. Member States shall ensure that a certification service provider may indicate in the qualified certificate a limit on the value of transactions for which the certificate can be used, provided that the limit is recognisable to third parties. The certification service provider shall not be liable for damage resulting from this maximum limit being exceeded. 5. The provisions of paragraphs 1 to 4 shall be without prejudice to Council Directive 93/13/EEC of 5 April 1993 on unfair terms in consumer contracts'.

[33] For an in-depth analysis of CSP liability as set forth in Article 6 of the Electronic Signatures Directive, see Balboni, P. (2004), pp. 211-242.

3.2 Model of adequate third-party liability for TMOs

On the basis of the results of the comparative analysis, it is concluded that, in theory, there seems to be enough legal ground in Europe to enforce TMO third-party liability towards e-consumers who rely on trustmarks and consequently suffer loss. However, in practice, the chances that TMOs will not be held liable accountable to e-consumers for the provision of inaccurate trustmarks are way bigger than the chances that TMOs will be held liable. Furthermore, considering TMO practice, the impact it has on e-consumers, e-merchants, governments, as well as on the same TMOs, and the deterring role that liability should have, being one of the first legal guards against undesirable societal behaviour,[34] it is concluded that the present TMO third-party liability system is not adequate (Chapter 8).

Therefore, a model of adequate third-party liability for TMOs will be elaborated. The core of TMO liability will be based on the same principles of surveyors', auditors', and accountants' liability which will be tailored to the specific TMO practice by developing the concept of 'adequacy'. More precisely, the concept of 'adequacy' is defined by applying to the trust relationship between TMOs and e-consumers the ethical theory of 'Warranted Trust' – which, in a nutshell, aims to protect trustor reliance on trustee by implementing a regulative framework which takes into consideration the interest of both parties and the influence of the specific context in which the trust relationship develops – and considering the social, economic, and political value of trustmarks.

There are two main reasons why an ethical theory is chosen in order to improve the TMO third-party liability system. First, this approach seems to widely comply with what has been recommended at the European level. In fact, in Recital 32 of the Directive 2000/31/EC of the European Parliament and of the Council of 8 June 2000 on Certain Legal Aspects of Information Society Services in the Internal Market,[35] the creation of codes of conduct

[34] 'Liability law has traditionally been the first legal guard against undesirable societal developments' Schellekens, M. & Prins, C. (2006), p. 49. In this respect, liability law aims to discourage subjects from putting in place specific unwanted conducts by making them liable for them and eventually obliging them to compensate the damages that may result from their misconduct.

[35] *OJ* L 178, 17.7.2000, p. 1.

at the Community level is encouraged as a means to determine the rules on professional ethics applicable to commercial communication.[36] As the present effort aims to contribute to the creation of European standards/code of conduct for TMOs, the ethical aspects of TMO practice need to be taken into consideration also in setting out the related liability system. Second, law and ethics are two normative sciences which traditionally can very well complement each other. To put it down very simply, ethics sets the basic societal interests that law should guarantee: '[w]ithin the tradition of natural law thinking which finds its roots in the philosophies of Aristotle and Aquinas, the political community has generally been understood in terms of a fundamental goal: that of fostering the ethical good of citizens. Law, on this concept, should seek to inculcate habits of good conduct, and should support a social environment which will encourage citizens to pursue worthy goals, and to lead valuable lives'[37] (Chapter 9).

---

[36] Recital 32: 'In order to remove barriers to the development of cross-border services within the Community which members of the regulated professions might offer on the Internet, it is necessary that compliance be guaranteed at Community level with professional rules aiming, in particular, to protect consumers or public health; codes of conduct at Community level would be the best means of determining the rules on professional ethics applicable to commercial communication; the drawing-up or, where appropriate, the adaptation of such rules should be encouraged without prejudice to the autonomy of professional bodies and associations.'

[37] Craig, E. (2005), p. 542. The complementary nature of law and ethics and their interaction will be clearly shown in Chapter 9, Section 2 *infra*.

Chapter 2
TRUST AND TRUSTMARKS

1. LACK OF E-CONSUMER TRUST IN E-COMMERCE

A quite recent survey of European citizens on issues relating to business and consumer e-commerce shows that 'lack of trust' in the Internet is the third most important reason for consumers to not buy online. The first reason is 'lack of access' (57% of the respondents), the second being 'lack of interest in buying anything using the Internet medium' (28% of the respondents); 25% of the respondent consumers said that they did not trust the Internet itself.[38] In other words, except for Internet access and consumer interest in buying something online, 'lack of trust' is indeed the first reason why consumers do not buy online.

Security, privacy, unfamiliarity with services, lack of direct interaction, and credibility of information seem to be at the top of the list of consumers' concerns in making online transactions.[39] If there is concern, trust is unlikely to develop. Therefore, if e-merchants want to gain consumers' trust, they have to start by reassuring consumers about the security of their website, the soundness of their privacy policy, and the reliability of the information provided by their websites. Moreover, e-merchants have to find ways to overcome potential clients' unfamiliarity with the services on offer. Last but not least, e-merchants have to deal with the absence of direct interaction with consumers, on the one hand, and the lack of consumers' direct interaction with the e-merchants' products, on the other hand.

---

[38] *Eurobarometer* (2004), pp. 9-11; see also Increasing Trust and Confidence of Consumers in the Information Society (forthcoming) Study for the European Commission. For the description of the project see <http://ec.europa.eu/information_society/eeurope/i2010/studies/studies_ongoing/index_en.htm>.

[39] See Egger, F. (2002), pp. 13-16; See also Organisation for Economic Co-operation and Development – Working Party on Indicators for the Information Society (2008), pp. 26 et seq.

2. TOP FIVE PSYCHOLOGICAL BARRIERS FOR E-CONSUMERS

In *Trust in Electronic Commerce,* Florian Egger points out security, privacy, unfamiliarity with services, lack of direct interaction, and credibility of information as the main psychological barriers to the adoption of e-commerce.[40]

Various statistics have shown that security is the biggest concern not only of e-consumers but also of e-merchants.[41] For example, security of payment is the leading concern of European citizens about buying through the Internet.[42] Moreover, bugs in (or breach of) e-merchants' security systems that allow hackers to steal passwords and all kinds of customers' personal information have often been reported.[43] Last but not least, the Consumer Sentinel database,[44] which is maintained by the Federal Trade Commission (henceforth: FTC) and which tracks both domestic and cross-border fraud, contains more than one million consumer complaints that can be classified in two categories: identity theft complaints and fraud complaints.[45] In summary, e-consumers do not feel secure in the online environment and thus are reluctant to provide payment details and personal data.

Part of the privacy concerns that have already been pointed out pertains to security. In fact, the two issues are closely connected.[46] However, e-

[40] See Egger, F. (2002), pp. 13-16; see also Organisation for Economic Co-operation and Development – Working Party on Indicators for the Information Society (2008), pp. 26 et seq.

[41] See Organisation for Economic Co-operation and Development – Working Party on Indicators for the Information Society (2005); see also Organisation for Economic Co-operation and Development – Working Party on Indicators for the Information Society (2008).

[42] 'In the context of issues of concern to those having used e-commerce, security of payment was still an important issue for 48% of EU15 respondents.' *Eurobarometer* (2004), p. 5.

[43] E.g., 'DSW reported on March 8, 2005 that credit card and personal shopping information on its customers was stolen from a corporate database over a three-month period' W.D.P.R. (2005) DSW Settles FTC Charges it Failed to Protect Sensitive Customer Data, *World Data Protection Report* 5 (12), p. 3.

[44] Available at <http://www.consumer.gov/sentinel/>.

[45] See also Organisation for Economic Co-operation and Development – Working Party on Indicators for the Information Society (2008), pp. 19 et seq.

[46] See Organisation for Economic Co-operation and Development – Working Party on Indicators for the Information Society (2008), p. 27.

consumer data are not always captured by malicious third parties who manage to bypass security systems. There are also e-merchants and dedicated organisations that intentionally collect e-consumer data.[47] A considerable number of users do not feel their privacy is sufficiently protected by the law of their countries.[48] Users are worried about leaving personal information (i.e., name, address, date of birth, gender) on the Internet.[49] They would like to know the reason why e-merchants or organisations gather their personal data and whether they share the data with other parties.[50] Actually, more than half of all European citizens are concerned about the broad issue of privacy protection.[51]

As to the unfamiliarity with the services, e-merchants have created many new services and business models. Perceived e-consumer risks exist in this new market. Novelty implies unfamiliarity which, coupled with perception of risk, is more likely to bring mistrust than trust. This can be explained by the lack of experience with such services, the lack of understanding of radical new business models, and, last but not least, lack of information on the e-merchants providing the new services. In fact, since a website can be created quickly, inexpensively, and at almost any location in the world, the e-consumer may question the very existence of the e-merchant. Furthermore, as the e-merchant might be unknown to e-consumers prior to his discovery on the Internet, concerns could arise about product quality, e-merchant reputation, and policies. E-consumers may also be worried about divulging confidential information, such as their credit card number, to an unknown e-merchant over the Internet.

The fact that commercial exchange is mediated via a computer screen or handsets may also impede the development of e-consumer trust in e-com-

---

[47] See, e.g., Listening to the Internet. Internet trends: Companies are eavesdropping on online discussion forums to find out what their customers really think about them, *The Economist*, 9 March 2006, p. 7.

[48] About 46% of the European citizens think that the level of data protection offered by the law in their own country is not high. See *Eurobarometer* (2003), p. 41.

[49] About 64% of the European citizens are worried about spreading personal information (i.e., name, address, date of birth, gender) on the Internet. See *Eurobarometer* (2003), p. 44. Only EU15 citizens were polled.

[50] Nine out of ten European citizens tend to agree that they should be informed why organisations are gathering personal data and if they are sharing it with other organisations. See *Eurobarometer* (2003), p. 40.

[51] See *Eurobarometer* (2003), p. 7.

merce.[52] There is a lack of direct interaction with people. Both salespeople and fellow shoppers can give clues about an e-merchant's trustworthiness in face-to-face interaction. On the Internet, salespeople are replaced by FAQs and search engines or are available only through electronic media.[53] The other kind of interaction missing online is obviously the interaction with the products themselves.[54] The lack of experimental interaction proves to be a particularly tough challenge for non-standard products, such as textiles or craft products. Despite sophisticated feedback devices, it will always be very difficult to adequately communicate factors associated with the intrinsic quality of a non-standard product through a computer interface.[55]

Lastly, since everybody can register a domain name and set up a website, it is sometimes hard to recognise websites of fair companies from those that are driven by malicious intentions. Sometimes, published information can be deliberately wrong or misleading. A common example is the case of allegedly objective product reviews that are sponsored by the manufacturer or fake testimonials. This problem is made worse by the fact that information can be altered simply and quickly, leaving no trace of the original text. Besides, websites are meant to be the most dynamic medium to date, as information can be updated and published instantly. However, this does not mean that all websites are always up-to-date, which can be a real problem in relation to prices, description of products or services, or, more generally, the availability of information. People appear to be quite aware of these issues. In fact, it is reported that one quarter of European citizens have had difficulties with the actual reliability of information on the Internet.[56]

3. Factors enhancing e-consumer concern

E-consumers' psychological barriers find further justification in the fact that e-commerce is marked by various new features, including de-

---

[52] See, e.g., Organisation for Economic Co-operation and Development – Working Party on Indicators for the Information Society (2008), p. 27.

[53] However, the anonymity of the seller is a cause of concern to only 16% of those who have bought something on the Internet. *Eurobarometer* (2004), p. 6.

[54] The main reason given by two thirds (68%) of the people who have never purchased online for lack of interest in buying on the Internet is that they need to see and touch the products they intend to buy. *Eurobarometer* (2004), p. 17.

[55] See Egger, F. (2002), p. 16.

[56] *Eurobarometer* (2004), p. 6.

materialisation, internationalisation or de-territorialisation, and technological turbulence.

The Internet has a de-materialising effect on identity, products, and information. Physical identity is replaced on the Internet by multiple virtual identities or by anonymity.[57] This makes it easier to carry out fraudulent actions (e.g., phishing[58] and pharming[59]). Furthermore, products are presented online to potential buyers by means of pictures, descriptions of the relevant qualities, and sometimes feedback from previous purchasers (i.e., people you do not know and who may not even exist). It is reported that people generally need not only see but also touch the products that they intend to buy.[60] Hence, de-materialisation of products represents a psychological barrier that e-merchants need to break down. Moreover, when the identity of a source of information is not known, or is simply not checkable, the credibility of the information is more difficult to assess. De-materialisation of information means that the information is no longer supplied in

---

[57] For an in-depth analysis of digital anonymity and the legal issues related, see Nicoll, C. et al. (2003); see also Ricci, A., Pelino, E., Balboni, P. and Cunegatti, B. (2008), pp. 281-344.

[58] 'In computing, phishing is a form of criminal activity using social engineering techniques, characterized by attempts to fraudulently acquire sensitive information, such as passwords and credit card details, by masquerading as a trustworthy person or business in an apparently official electronic communication, such as an e-mail or an instant message. The term phishing arises from the use of increasingly sophisticated lures to fish for users' financial information and passwords.' *Wikipedia*. See <http://en.wikipedia.org/wiki/Phishing>.

[59] 'Pharming is a hacker's attack aiming to redirect a website's traffic to another (bogus) website. Pharming can be conducted either by changing the host's file on a victim's computer or by exploitation of a vulnerability in DNS server software. DNS servers are computers responsible for resolving Internet names into their real addresses — they are the 'signposts' of the Internet. Compromised DNS servers are sometimes referred to as 'poisoned'. The term pharming is a word play on farming and phishing. The term phishing refers to social engineering attacks to obtain access to credentials such as user names and passwords. In recent years both pharming and phishing have been used to steal identity information. Pharming has become of major concern to businesses hosting ecommerce and online banking websites. Sophisticated measures known as anti-pharming are required to protect against this serious threat. Antivirus software and spyware removal software cannot protect against pharming. Pharming is becoming the attack du jour of today's hackers.' See <http://en.wikipedia.org/wiki/Pharming>.

[60] *Eurobarometer* (2004), p. 17. As already mentioned in n. 54, the survey shows that 68% of the people who have never bought anything online maintain that the main reason for it is that they need to see and touch the product first.

written form but electronically. Files, software, and other information are no longer made available using physical carriers but are increasingly being made available online for downloading purposes. This third effect of dematerialisation has a significant impact on security and privacy protection. In fact, on the one hand, e-merchants have to secure invisible data flows (e.g., transactions, e-consumers' personal data communications). On the other hand, e-consumers are unable to check how their dematerialised data are used by e-merchants.[61]

Internationalisation is the change brought about by the Internet as a means of communication not restricted by geographic borders. It will be extremely challenging, due to the mobility of the traffic on the Internet and the tendency of senders to hide their identity, to determine where exactly a particular person is or where an activity is being carried out. Moreover, it is becoming more and more difficult to determine which route an electronic message has taken before it eventually reaches its destination. It may be true that the borderless nature of the Internet gives e-consumers access to a great amount of information; but this also makes it extremely complex for them to verify the source of the information and to have a direct interaction with the person behind the source. Internationalisation has given rise to worldwide competition in lower prices for e-consumers. However, rules (e.g., consumer or data protection rules) are not the same everywhere, so e-consumers may be more at risk in some places.

Technological turbulence refers to the constant and rapid development of technology. One could say that the Internet brings to the e-consumer his daily (technological) challenge. In fact, owing to the dizzying evolution of technology, new services or updates of the existing ones are offered online almost everyday. The upside of this phenomenon is that e-consumers are constantly presented with better quality of services. The downside is that the same technological improvement may also be used for malicious purposes. For example, more sophisticated fraud can be carried out online, more powerful viruses[62] can be sent to harm user computers, and more

[61] For a more detailed analysis of 'de-materialisation' and the related legal issues, see De Cock Buning, M. et al. (2001), pp. 291 et seq; Van Klink, B.M.J. & Prins, J.E.J. (2002), pp. 9-10.

[62] 'A computer virus is a self-replicating computer program written to alter the way a computer operates, without the permission or knowledge of the user. Though the term is commonly used to refer to a range of malware, a true virus must replicate itself, and must

advanced spyware[63] can be used to collect user preferences. Technology itself already represents an obstacle for many people. They have to become familiar with the technology and learn how to use it. However, it takes a lot of effort to become familiar with something that constantly evolves. If the people do not feel that they are familiar with the situation, they feel more vulnerable and develop concerns about online security.

4. Information asymmetry as the key issue

De-materialisation, internationalisation, and technological turbulence create a great variety of information, products, and services for e-consumers. However, the source of the information and the quality of the products and services are not easily verifiable by e-consumers. Generally, e-consumers do not have enough competence to master the technology on which e-commerce is based. This causes far-reaching information deficits on the e-consumers' side. Therefore, it may be argued that the very reason for the consumers' psychological barriers mentioned above can be traced to information asymmetry (see Table 3).

Generally, information asymmetry exists when one party to a transaction has more or more accurate information than the other party (usually it is the seller that knows more about the product than the buyer).[64] Information asymmetry is not a specific feature of the online market, offline mar-

---

execute itself. The latter criteria is often met by a virus which replaces existing executable files with a virus-infected copy. While viruses can be intentionally destructive – destroying data, for example – some viruses are benign or merely annoying.' See <http://en.wikipedia.org/wiki/Computer_virus>.

[63] 'In the field of computing, the term spyware refers to a broad category of malicious software designed to intercept or take partial control of a computer's operation without the informed consent of that machine's owner or legitimate user. While the term taken literally suggests software that surreptitiously monitors the user, it has come to refer more broadly to software that subverts the computer's operation for the benefit of a third party. In simpler terms, spyware is a type of program that watches what users do with their computer and then sends that information over the internet. Spyware can collect many different types of information about a user. More benign programs can attempt to track what types of websites a user visits and send this information to an advertising agency. More malicious versions can try to record what a user types to try to intercept passwords or credit card numbers. Yet other versions simply launch popup advertisements.' See <http://en.wikipedia.org/wiki/Spyware>.

[64] See, e.g., Arrow, K.J. (1963), pp. 141-149, where this situation was first described.

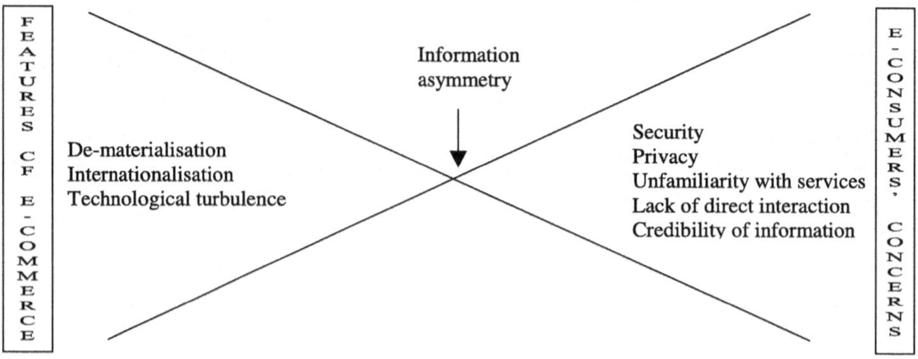

Table 3. Information asymmetry

kets are also often characterised by far-reaching information deficits that impede smooth functioning.[65] Quality labels, as the result of a neutral certification process,[66] have been used in offline markets to reduce psychological barriers.

Considering the European food sector, for example, Gabriele Jahn et al. in their paper *The Reliability of Certification: Quality Labels as a Consumer Policy Tool* aptly describe the origin and the value of quality labels in the food sector.[67] They argue that quality labels based on neutral control throughout the value chain have been adopted to enhance consumer protection.[68] In fact, consumers' confidence in the safety and quality of food products has decreased after the recent scandals (e.g., BSE, FMD, and bird flu), which abundant regulation and government controls were not able to prevent.[69] Concerning quality labels as a policy tool, the authors point out that, on the one hand, by outsourcing labelling services, public authorities are relieved of a financial burden. On the other hand,[70] the reliability of the

[65] See Akerlof, G.A. (1970), pp. 488-500. See also Spence, M. (1976), pp. 591-597.

[66] 'Certification is a procedure by which a third party gives a written assurance that a product, process or service conforms to specific characteristics', Rae, A. et al. (1995), p. 2.

[67] See Jahn, G. et al. (2005), pp. 53-73.

[68] Id., p. 53.

[69] See also Hobbs, J.E. et al. (2002), pp. 77-81; Sporleder, T.L. & Goldsmith, P.D. (2001), pp. 591-604.

[70] See also Caswell, J.A. & Mojduszka, E.M. (1996), pp. 1248-1253.

quality labels and their effectiveness in consumer policy strongly depend on the quality of the assessment carried out.[71]

Food quality certification has taught three important lessons. First, regulations and governmental control have proved inadequate to guarantee the safety and quality of (food) products. Second, third-party certification (i.e., quality labelling) has been used by European countries as a policy tool to improve consumer protection, win back consumer trust, and, at the same time, relieve public authorities of an additional financial burden. Third, the central task of certification, i.e., the reduction of information asymmetry within the market, can be fulfilled only if the institutions in charge succeed in assuring certification quality and thus the validity of the audit signal.

Another important lesson can be learned from the financial market. In this market, third-party audits are carried out to detect and prevent errors, fraud, and lack of conformity to external and internal rules and standards. In fact, external auditors are required to examine the books, vouchers, records, and accounts of a company with a view to ascertaining whether they represent a true and fair view of the company's affairs, whether they have been prepared in conformity with generally accepted accounting principles, and whether these principles have been consistently applied over the years. However, the recent crisis in financial auditing reflects the potential shortcomings of third-party control procedures. Enron and Parmalat were two scandals which, to use a euphemism, significantly diminished confidence in the quality of financial auditing.[72] The lesson to learn here is that the reputation of the auditing company is crucial to the certification business. At this point, it has to be borne in mind that '[a]ny player in the market suffers reputation damage when there is a failure of one sort or another.'[73]

---

[71] 'The control process is usually carried out by independent inspectors (certifiers) who, in their turn have to meet criteria laid down by rule-making agencies. Only if the certifiers succeed in revealing critical aspects and opportunistic behaviour will quality assurance concepts be able to build up the reputation necessary to serve as a reliable quality signal.' Jahn, G. et al. (2005), p. 53.

[72] See Nussbaum, B. (2002), pp. 39-55; Thomas, C.W. (2002), pp. 41-48; Vinten, G. (2003), pp. 445-448.

[73] Samuel DiPiazza, global head of PricewaterhouseCoopers LLP, in an interview with the *Financial Times* on 21 July 2005 quoted in Parker, A. (2005), p. 11.

5. Good news: trustmarks

As in the offline market, third-party certification has also been used to reduce information asymmetry in the online market. In the late 1990s, a potential solution to the need for trust of consumers and merchants was offered in the US by private companies called trustmark organisations (TMOs). In a nutshell, TMOs are organisations which present themselves as independent parties that provide trustmarks – labels or visual representations indicating that a product, process, or service conforms to specific quality characteristics – to e-merchants. Security, privacy, and business practice[74] are three fields in which trustmarks are popular. In fact, a TMO will issue a trustmark to an e-merchant if it has demonstrated that it conforms to the policy of the TMO regarding security, privacy or business practice. The e-merchant hopes that, by displaying the trustmark on its website, an e-consumer will less likely question the integrity of the e-merchant in relation to the security, privacy or business practice. Moreover, the trustmark, as a third-party professional guarantee for various e-merchant activities, addresses e-consumer concerns, such as unfamiliarity with services, lack of direct interaction, and credibility of information. Thus, the e-consumer will more likely divulge his personal data and transact with the e-merchant, i.e., conduct online business. In summary, TMOs are independent organisations that try to promote online trust by offering a system of certification. VeriSign,[75] TRUSTe,[76] BBBOnline,[77] and WebTrust[78] are the most popular TMOs in the US.

The nature and complexity of the Internet encourage the use of institution-based assurances because they are a signal to the users of legitimacy and trustworthiness. TMOs offering e-commerce assurance services may be perceived in the electronic marketplace as third-party endorsers. Research on marketing literature shows that third-party endorsements function similarly to expert endorsements.[79] In other words, consumers may perceive TMOs as experts/professionals who control and eventually guarantee the conduct of e-merchants.

---

74 'Business practice' is a term which has a very broad meaning and embraces various e-merchant practices, e.g., security, privacy, marketing, information.

75 See <http://www.verisign.com/>.

76 See <http://www.truste.org/>.

77 See <http://www.bbbonline.org/>.

78 See <http://www.webtrust.org/>.

79 See Dean, H. & Biswas, A. (2001), pp. 41-57.

When there is a trusting attitude, consumers are more comfortable in sharing personal information and purchasing online.[80] According to studies carried out in the US, there seems to be a positive relationship amongst institution-based structures (e.g., trustmarks), online trust, and intent to purchase.[81] For example, in a survey conducted by Harris Interactive,[82] on behalf of VeriSign, in September 2004, it was reported that 74% of people in the US who have ever made an online purchase look for a trustmark when determining whether or not to buy from an e-merchant website.[83]

In Europe, the trustmark phenomenon is still in its infancy. Confianza Online,[84] Euro-Label,[85] Luxembourg e-Commerce Certified,[86] Thuiswinkel,[87] and Trusted Shops[88] are some of the European TMOs. What was pointed out in 2003 by Ton Wagemans in his paper *An introduction to the labelling of Websites* is still valid: '[whether] the existing labelling initiatives really have increased consumer trust in internet and e-business is hard to measure, because most of them are still in a development stage and because of the lack of data.'[89] Only one in ten European citizens in 2004 had heard of Internet trustmarks.[90] However, in a recent study on trustmarks in Europe, it has been reported that e-consumer representatives foresee that the largest effect of a trustmark scheme will be the increased willingness of consumers to leave personal information on a website of a certified e-merchant. In terms of online shopping, consumer representatives feel it is more likely that consumers will buy online more often and at different shops.[91]

---

80 See McKnight, H. et al. (2002), pp. 334-359.

81 See Lala, V. et al. (2002), pp. 237-253; Odom, M. et al. (2002), pp. 231-250. Empirical research also relates the positive effect of third-party certification on the likelihood of online purchases and trusting attitudes, see Mauldin, E. & Arunachalam, V. (2001), pp. 33-54.

82 <http://www.harrisinteractive.com/>.

83 See Organisation for Economic Co-operation and Development – Working Party on Indicators for the Information Society (2005), p. 14.

84 See <http://www.confianzaonline.org>.

85 See <http://www.euro-label.com/euro-label/ControllerServlet>.

86 See <http://www.e-certification.lu/index.html>.

87 See <http://www.thuiswinkel.org/>.

88 See <http://www.trustedshops.de/de/home/index.html>.

89 Wagemans, T. (2003), p. 2.

90 See *Eurobarometer* (2004), p. 20.

91 See De Bruin, R. et al. (2005), Final Report, p. 32.

5.1 Trustmark benefits

Trustmarks are a very valuable means to improve e-society and e-economy and to simplify e-policy. In fact, not only e-consumers but also e-merchants and governments can benefit a great deal from trustmarks.

E-consumers are not able to scrutinise the policies of organisations, companies, and other participants of the virtual world without borders.[92] Ideally, through trustmarks, e-consumers can receive a sort of guarantee from an independent third party of the quality of, for example, the e-merchants' business practice, their privacy statement, or the security level of their websites. Moreover, trustmarks are very easy to recognise and can improve the perception of e-consumers with regard to potential online business partners, provide 'always-available', independent, and trustworthy information on e-merchants and thus enhance e-consumer confidence in online transactions.

If the benefits of trustmarks to e-consumers can be summarised by the concept of a better buying experience, the consequential benefit to e-merchants will be a better selling experience. Given the big pressure to which merchants are exposed in the online world, which has been neatly described by Assafa Endeshaw in his paper *The Legal Significance of Trustmarks*, as follows: '[w]hile it has taken many international companies a succession of decades to establish trust and confidence in the line of business or type of wares and services they have offered to their customers, the current wave of business start-ups on the Internet feel hard pressed to succeed fast in the "Wild Web" lest they perishes with the same speed that they sprouted',[93] trustmarks can help e-merchants succeed fast. Through the exhibition of the trustmark on their websites, e-merchants can make some information easily available to e-consumers, increase the chance to win their trust, and eventually do some business with them. Moreover, trustmarks offer e-merchants the chance to self-regulate sectors of their activities, set out their own standards, and thus prevent governments from interfering.

Last but not least, enhancing consumer trust in e-commerce tops the European Union agenda.[94] The reason is very simple: nowadays, trust is

---

[92] See Section 3 *supra*.

[93] Endeshaw, A. (2001), p. 204.

[94] See, e.g., the recent study for the European Commission Increasing Trust and Confidence of Consumers in the Information Society (forthcoming); *Eurobarometer* (2004); i2010

money,[95] and with a minority of the European citizens engaged in e-commerce activities,[96] there is still a long way to go in terms of reaping the full benefits of e-commerce. Lack of trust is the first reason why consumers do not buy online, given the presence of Internet access and interest in purchasing anything online.[97] Trustmarks are a means of self-regulation which aims to enhance the e-consumers' trust in online communications. Therefore, trustmarks can give governments the chance to stay out of some sectors of Internet regulation, thus relieving them of an additional financial burden. Concurrently, trustmarks can help boost e-commerce and bring governments the related revenues.

5.2 Reputation: TMO success key

TMO reputation and credibility are key elements for the success of trustmarks. Without them, the system will not take off. Robin Wakefield and Dwayne Whitten, in their paper *Examining User Perceptions of Third-Party Organization Credibility and Trust in an E-Retailer*, proved that TMO credibility is positively related to e-consumer trust in e-merchants.[98] They demonstrated their statement in two steps. First, the authors showed that TMO credibility is positively related to the value of a trustmark (i.e., web assurance seal) and that the value of a trustmark is positively related to trust in the e-merchant.[99] Second, they proved that the value which Internet users assign to assurance structures is negatively related to perceptions of purchase risk and that the perception of purchase risk is negatively related to trust in the e-merchant.[100]

Such finding matches earlier results of signalling literature, according to which signals tend to be effective when the signalling firm (e.g., a TMO)

Five-year strategy to boost digital economy <http://europa.eu.int/information_society/eeurope/i2010/index_en.htm>; the setup of a dedicated European Agency on Information Security (ENISA) also highlights the political significance of information security and the need to strive for greater cooperation across EU Member States as well as internationally, see <http://www.enisa.eu.int/>; see also the very recent document produced by the ENISA (July 2007).

95 See, e.g., Valente, K. (2001).

96 *Eurobarometer* (2004), p. 3.

97 *Eurobarometer* (2004), pp. 9-11.

98 See Wakefield, R.L. & Whitten, D. (2006), pp. 1-19.

99 See id. pp. 6-7.

100 See id. p. 7.

maintains a high reputation.[101] Furthermore, organisational credibility plays a role in influencing attitudes and purchase intentions[102] and is an important component of reputation.[103] Ultimately, trust may have less to do with privacy, security, and business practice and more to do with the reputation of the TMO.[104]

This has been indirectly confirmed also by Fredrik Nordquist et al. in their paper *Trusting the Trustmark?*, in which they point out that 'the average e-consumer does not possess the necessary knowledge or eagerness to evaluate all existing TMs [trustmarks] and CoCs [Codes of Conduct]. He will therefore find himself in a position of ignorance in regards to the actual implications of the individual TMs for him as a consumer. He will therefore, according to the theory of asymmetric information,[[105]] consider all TMs to be of equal quality.'[106] In fact, e-consumers are unaware of the details of the trustmark process. They do not know where to find this information. If they find it at all, they will most likely not understand it, and it is a time-consuming exercise, anyway.

Hence, e-consumers basically rely on TMO reputation.

Such statement has been confirmed also by Bernard Brun, who wrote in his paper *Nature et impacts juridiques de la certification dans le commerce électronique sur Internet*: 'Dans la réalité, il ne sera pas question de reconnaissance de la certification mais plutôt de reconnaissance du certificateur'.[107]

6. Bad news: TMO practice has already been proven to be untrustworthy

6.1 A case study

Despite the fact that online quality certification is still in its early stages, the unpleasant events – not to say scandals – in which TRUSTe was involved in

---

[101] See Hoxmeier, J. (2000), pp. 115-139.
[102] See Goldsmith, R. et al. (2000), pp. 43-54; Lafferty, B. et al. (2002), pp. 1-12.
[103] Keller, K.L. (1998).
[104] See Wakefield, R.L. & Whitten, D. (2006), p. 1.
[105] See Akerlof, G.A. (1970).
[106] Nordquist, F. et al. (2002), pp. 5-6.
[107] Brun, B. (2001), p. 46.

1999 and 2000 show some weaknesses of the US trustmark system.[108] A brief description of what happened is provided hereunder.

In March 1999, Microsoft (which was TRUSTe-certified) was caught transmitting user information through the Windows 98 registration wizard. However, TRUSTe refused to act decisively because the breach did not involve a website but a software, which was not covered in its charter.

In September 1999, a security hole in Microsoft's web-based e-mail programme Hotmail allowed hackers to snatch users' IDs and passwords. Microsoft fixed the bugs and, at TRUSTe's request, submitted to a third-party privacy audit.

In November 1999, a news report revealed that RealJukebox, a software plug-in by RealNetworks, was surreptitiously collecting data on users' music-listening habits and passing them back to RealNetworks. Working with TRUSTe, RealNetworks discontinued the practice, appointed a privacy compliance officer, and checked its privacy policies. However, as in the first case reported, because the violation involved a piece of software, which was not covered in its charter, TRUSTe declined to take further action.[109]

Moreover, as Michael Froomkin stated in his paper *Death of Privacy?*: '[T]he RealNetworks incident followed an earlier, similar fiasco',[110] referring to a case in which the FTC settled a complaint against GeoCities.[111] In fact, the FTC discovered that GeoCities had collected personal identifying information from its users (i.e., children and adults) in order to create a database that included e-mail and postal addresses, member interest areas, and demographics, including income, education, gender, marital status, and occupation, and eventually disclosed customer data to marketing companies without adequately stating the purposes of such data processing.[112] However, GeoCities' privacy statement set forth that the customers' registration information would be used only to 'provide members the specific advertising offers and products or services they requested and that the

---

[108] See Froomkin, M. (2000), p. 1526.

[109] For more information on the case, see Open Letter 1999/11/1 to RealNetworks; Lemos, R. (1999) Can You Trust TRUSTe?, *ZDNet News*, 2 November 1999. Available at <http://news.zdnet.com/2100-9595_22-516377.html>.

[110] See Froomkin, M. (2000), pp. 1526-1527.

[111] See McCarthy, J. (1999). The article details several other debacles, in which trustmark holders violated privacy policies or principles but kept their accreditation.

[112] See Kornblum, J. (1998).

"optional" information [education level, income, marital status, occupation, and interests] would not be released to anyone without the member's permission.'[113] Accordingly, the FTC maintained first that GeoCities provided its customers with a misrepresentation of the purposes for which their personal data were processed and, second, that user data communication occurred without prior data subject permission/consent. GeoCities settled the complaint, although it denied the allegations. Concurrently, it modified its privacy policy by clearly stating that subject to prior user consent, user data might be disclosed to third parties. It has to be noted that during the FTC investigation, TRUSTe's trustmark was not removed from GeoCities' website.[114]

On 1 February 2000, a healthcare group revealed breaches of the stated privacy policies of 16 healthcare websites, including seven TRUSTe members (i.e., AltaVista, CBS.com, Excite.com, Healtheon/WebMD, Mediconsult, MotherNature.com, and Yahoo). Among other things, the sites were criticised for failing to disclose that advertising banner networks they participate in, such as DoubleClick, let advertisers collect any personal information disclosed by a customer on a page where an advertisement appeared. A week after the incident, TRUSTe stated that it was working with licensees to ensure compliance with the programme, although TRUSTe did not take any further action against the e-merchants' misconduct.

Finally, on 7 February 2000, a news report revealed that, while E-Loan did not use cookies or share users' personal data with third parties, several recently acquired subsidiaries did, violating E-Loan's stated policy. The firm quickly posted a rewritten policy on its website. Because E-Loan acted promptly to align its policy and practice, TRUSTe took no action.

These examples have some common features. First, the privacy programme of TRUSTe was violated. Second, TRUSTe was seemingly not aware of it until the violation was pointed out by some outsider (a news report in the E-Loan case, a complaint by a health care group in the second example, a news report in the RealJukebox case, etc.). Third, none of the trustmarks was revoked. In fact, what TRUSTe did in some cases was to work with the e-merchant to restore compliance with its programme (e.g., healthcare companies case, RealJukebox case, and Hotmail case); in

[113] Kornblum, J. (1998) (quoting GeoCities' membership sign-up form).
[114] See McCarthy, J. (1998).

the E-Loan case, it was the e-merchant, on its own initiative, which discontinued the practice, and in the Microsoft case, TRUSTe did not take any action at all, denying its responsibility for the issue at stake. Consumers' data were kept, shared, or sold by e-merchants which had TRUSTe's trustmark on their websites, without TRUSTe even being aware of it. Nonetheless, TRUSTe's trustmark was still on the e-merchants' website after the violation was discovered.

It is argued by some American doctrine that the reasons for these incidents can be traced to TRUSTe's funding structure and to the US self-regulation policy.[115] TRUSTe (as with other TMOs) receives funds from big high-tech companies, which, in some cases, also happen to be the same e-merchants that pay the yearly fee for the licence to use the trustmarks and whose representatives also sit in the TRUSTe board of directors. As a result, TRUSTe will think twice before denouncing the non-compliance of an e-merchant with its trustmark programmes and eventually revoking their respective trustmarks. TRUSTe certainly has no economic incentive to be tough on its funding sources.[116] On the other hand, the US self-regulation policy does not offer legal sanctions to incentivise or enforce strong privacy compliance[117] in a marketplace where the economic incentive to provide strong privacy protection is either weak, nonexistent, or at least not uniformly distributed amongst all the participants.[118] Roger Clarke in his paper *The Legal Context of Privacy-Enhancing and Privacy-Sympathetic Technologies* interestingly portrayed the consequence of this policy as follows: 'Wolves self-regulate for the good of themselves and the pack, not the deer';[119] and the TRUSTe scandals are cases in point.

6.2 Marketing-based trust

As already pointed out in Subsection 5.2, e-consumers ignore the practice behind a trustmark (i.e., the code of conduct, the assessment system, moni-

---

[115] See, e.g., Farrell, H. (2005); Froomkin, M. (2000), pp. 1525-1529; Regoli, N.L. (2002), pp. 369-370; Rafter, M. (2000).

[116] See Froomkin, M. (2000), p. 1527. This issue, which pertains to several TMOs, will be extensively dealt with in Chapter 3, Subsection 5.1 *infra*.

[117] The US may be unique in endorsing self-regulation without legal sanctions to incentivise or enforce it, see Clarke, R. (7 July 1998).

[118] See Froomkin, M. (2000), p. 1525 and p. 1528.

[119] Clarke, R. (12 April 1999).

toring activities, and the enforcement system). They basically rely on the reputation of TMOs. Therefore, TMOs benefit more from investing money in advertising their brand than in improving the quality of their trustmark system (i.e., a sound code of conduct, an independent assessment system, proactive monitoring of the compliance of an e-merchant, and a strong enforcement system). It is obvious that a trustmark system which is based more on marketing than on quality can be very dangerous for e-consumers and, in the long run, for the credibility of the trustmark system itself. A typical example of this practice and its consequences is again offered by TRUSTe.

In 2000, Jens Riegelsberger and Martina Angela Sasse wrote in their paper *Trust me, I'm a .com. The Problem of Reassuring Shoppers in Electronic Retail Environments*: 'Surprisingly, one of the leading advertisers on the Internet in the past half-year has been TRUSTe, an organisation that assigns seals to e-commerce enterprises which it considers "trustworthy". In order to raise its profile, TRUSTe has instigated an internet-based publicity campaign involving between 200 and 700 million weekly impressions for its adverts. "Why?" The number of B2C e-commerce retailing activities has not reached the dizzying heights predicted 18 months ago. Consumers' "lack of trust" in e-commerce is widely assumed to be one of the main reasons.' [120] The authors should not have been surprised to find TMOs amongst the leading advertisers on the Internet. These figures show the importance of marketing a brand in the certification process. However, TRUSTe tried to win e-consumers' trust in order to build its reputation by a marketing operation, although offering a service not up to expectations. More generally, it is hardly surprising that a company which invests a major part of its budget in marketing operations to catch e-consumers' attention will not have much money left to provide a high-quality service. A marketing-based reputation represents a big risk to e-consumers who rely on TMO practices. Furthermore, this is potentially very dangerous for the trustmark system as a whole because, in the certification business, '[a]ny player in the market suffers reputational damage when there is a failure of one sort or another.'[121] It is easier and more effective to do things right

[120] Riegelsberger, J. & Sasse, M.A. (2000).

[121] Samuel DiPiazza, global head of PricewaterhouseCoopers LLP, in an interview with the *Financial Times* on 21 July 2005 quoted in Parker, A. (2005), p. 11.

from the very beginning than be forced by scandals to have to win back consumers' trust. This will be confirmed by the four big accounting firms that are nowadays, after the Enron scandal, struggling to restore trust in the accountancy profession.

7. European TMOs' business model

Turning attention now to European TMOs, it is relevant to check whether the American TMO practice has also been implemented by European TMOs.

In a recent study on trustmarks, it is reported that there are seven critical success factors according to European TMOs:

1. Awareness of business and consumers
2. A detailed and robust code of conduct
3. Effective enforcement mechanisms
4. A sufficient number of trustmarks issued (leading to user fee revenue)
5. Trust in the (independent) organisation that operates the TMO
6. Stakeholder support
7. Low up-front and operational costs [122]

Awareness of business and consumers, which could be summarised in one word – 'marketing' (critical success factor no. 1)[123] – has already been

[122] '[M]ost respondents mentioned that raising awareness amongst companies and consumers is a critical success factor. A highly elaborated and robust code of conduct that defines criteria exceeding the strict legal requirements and provides measurable proven legal rights increases the perceived value of the trustmark scheme. Another critical success factor is the existence of effective enforcement mechanisms, such as audits (including helping applicants to meet the criteria), monitoring and an independent alternative dispute-resolution system. The number of trustmarks issued (leading to user fee revenue) and in which pricing plays a role is critical. Also mentioned is the trust from both businesses and consumers in the (independent) organisation that operates the trustmark scheme. A critical success factor that is closely related to this is the support of stakeholders, including government. Finally, low up-front and operational costs were indicated as critical success factor.' De Bruin, R. et al. (2005) Final Report, p. 97.

[123] Marketing is estimated to amount to as much as 23% of the total costs of the start-up phase. De Bruin's study acknowledges that a TMO's success depends on its visibility with business as well as consumers. Therefore, it is necessary for its logo to have a strong brand. High brand recognition requires communication through various media channels and there-

proven to be extremely important in American TMO practice[124] and is also a prominent element of the European TMOs' business model. However, TMOs declare to also pay a great deal of attention to developing a detailed and robust code of conduct (critical success factor no. 2) and a strong enforcement mechanism (critical success factor no. 3) which guarantees a fair audit, effective monitoring, and an independent alternative dispute resolution system.

At this point, a fine observer could ask himself how European TMO are able to combine a detailed and robust code of conduct and a strong enforcement mechanism with low up-front and operational costs (critical success factor no. 7). The answer to such question rests in private sponsorship (critical success factor no. 6). Private sponsorship, in the form of investment of the funding organisation(s) or future trustmark users, is an important source of money for performing TMO services. As it has been pointed out in the analysis of the case study, however, this direct financial link between the TMO and its members may imply a conflict of interest, notably when sanctions have to be enforced.[125]

Moreover, it is enough to dig a bit deeper into the European TMOs' practice to see a few differences between what TMOs point out to be critical success factors and the service they actually offer. Regarding the effectiveness of the monitoring, most TMOs do not seem to put in place all the possible mechanisms for establishing and monitoring compliance with their specifications.[126] Furthermore, as will be discussed in the next chapter, almost none of the TMOs analysed has a full-scale enforcement structure in place.[127] One of the reasons for this can be related to the costs of such a mechanism. Furthermore, for reasons of commercial viability, TMOs prefer quantity (number of clients) over quality (effective enforcement infrastructure) in the start-up phase.[128]

---

fore can be very costly. Furthermore, marketing expenses have to be seen as permanent costs. Once the brand has been created, it requires permanent marketing investment to retain or increase brand recognition. Cutting down on marketing, means less visibility and thus less brand recognition.

[124] See Subsection 6.2 *supra*.

[125] See Subsection 6.1 *supra*.

[126] See De Bruin, R. et al. (2005) Final Report, p. 78.

[127] See Chapter 3, Subsection 5.4 *infra*.

[128] See De Bruin, R. et al. (2005) Final Report, p. 103.

In summary, the practice of European TMOs does not differ much from that of American TMOs. As the American TMOs' practice represents a risk to e-consumers who possibly rely on trustmarks, the European one does so, too. Furthermore, given the present practice, scandals such as the ones in which TRUSTe was involved are likely to happen in Europe. This will have a devastating effect on the European TMOs' reputation and credibility of European TMOs and, ultimately, on the e-consumers' trust in e-commerce.

8. A LOSE-LOSE SITUATION

The TMO business consists of addressing e-consumer concerns by providing assurance services to e-merchants. A TMO's goal is to promote trusting attitudes that will allow electronic transactions to proceed smoothly. E-merchants that display a TMO's web assurance seal (i.e., trustmark) communicate to the consumer certain affirmations concerning legitimacy, security of transactions, privacy, and integrity. Research shows that web seals promote feelings of security and trust[129] and influence the e-consumers' intent to purchase online.[130] In other words, TMOs help e-merchants obtain e-consumer data and sell goods and services.

However, TMOs have already been proven to be not very trustworthy and, at first glance, their present practice does not look very promising. Risks of undetected e-merchants' actions that can cause loss to e-consumers are high, and damages can be difficult to recover.[131]

Not only e-consumers but also e-merchants and governments will lose as a result of an untrustworthy TMO practice.

By joining a trustmark programme, e-merchants run the risk that other e-merchants with the same trustmark will violate the rules of the trustmark programme and eventually cause damage to e-consumers who happen to

---

[129] See Houston, R. & Taylor, G. (1999), pp. 89-105; Palmer, J. et al. (2000).

[130] See Kovar, S. et al. (2000), pp. 17-35; Mauldin, E. & Arunachalam, V. (2001), pp. 33-54; Wakefield, R.L. (2001).

[131] Damages were quite difficult to quantify in, e.g., the two class action lawsuits filed on a federal level, the other in California for the surreptitious harvesting of music preference data by RealJukebox. In the federal case, the plaintiffs sought a refund of the 30 dollars that some users paid for the registered version of the software. In the California case, plaintiffs' lawyers very optimisticly estimated in 500 dollars per person the value of music preference data. See McWilliams, B. (November 10, 1999); Molland, M. (2001), pp. 197-200.

rely on that trustmark. In this way, the reputation of that trustmark programme will be damaged and the investments made by all e-merchants to join it will be wasted. In fact, this is a risk that will always exist, but an unreliable TMO practice will increase it exponentially. Damage to e-consumers caused by a certified e-merchant will reduce the already scarce trust that e-consumers have in e-commerce. This will not bring any positive effect to e-business or to the economy. Moreover, governments run the extra risk of letting TMOs run wild. In fact, if damage related to e-consumers' reliance on trustmarks occurs, the inactivity of governments in regulating the matter will also be indicated as one of the reasons for the TMOs' unreliable practice.

From this quick review of TMO practice, it seems that TMOs are heading in the wrong direction. However, this is only a provisional conclusion. It is now necessary to undertake an in-depth analysis of how the certification system offered by TMOs works, in order to check the grounding of this preliminary assumption.

Chapter 3
TMO PRACTICE

1. A DEFINITION OF CERTIFICATION

The word 'certification' derives from the Latin adjective *certus*, which means 'determined, resolved, fixed, settled, purposed'.[132] In fact, the most common perception of certification is that it gives some form of guarantee, generally of quality and dependability in their widest sense. Such perception could find its justification in the certification procedure according to which 'a third party gives a written assurance that a product, process, or service conforms to specific characteristics'.[133] The key element in the certification process is indeed the third party, an independent party who is expected to give an assurance (a guarantee) of the qualities of some products or services through the issuance of a certificate.

Moreover, Jean-Marie Ponthier, in his paper *La certification outil de la modernité normative*, compares certification with scales as they can both be used in a comparative analysis.[134] In fact, certification is mostly a means to measure products and services based on (quality) criteria which are established in advance. In this way, certification facilitates the comparison between different products and services.

The importance of product and service quality certification schemes for international markets is evident. In fact, when goods and services are exchanged worldwide, most of the time, their quality cannot be directly tested. An independent third-party guarantee, given through the issuance of quality certificates, is therefore very useful. E-commerce is actually the ulti-

---

132 Lewis, C.T. (1996) *A Latin Dictionary* (New York: Oxford University Press), p. 320.
133 Rae, A. et al. (1995), p. 2.
134 Ponthier, J. (1996), p. 355.

P. Balboni, Trustmarks in E-Commerce

mate expression of market internationalisation,[135] so certification can play an essential role there.[136]

2. The certification process

Certification is a process in which five stages can be distinguished (see Table 4). Two stages are in the pre-certification phase, i.e., before the issuance of the certificate. The third stage is indeed the possible issuance of the certificate. The other two stages are in the post-certification phase, i.e., after the certificate has been granted. The process unfolds as follows:

1. The standards need to be set.
2. An assessment of the entity to be certified has to be carried out.
3. Depending on the positive or negative outcome of the evaluation, the certificate will be issued or denied.
4. In the post-certification phase, the certificate has to be monitored.
5. Depending on the positive or negative outcome of the monitoring procedure, it may be confirmed or revoked. The certificate may also be suspended pending further enquiries, or to give the certified entity the chance to restore its compliance with the standards.

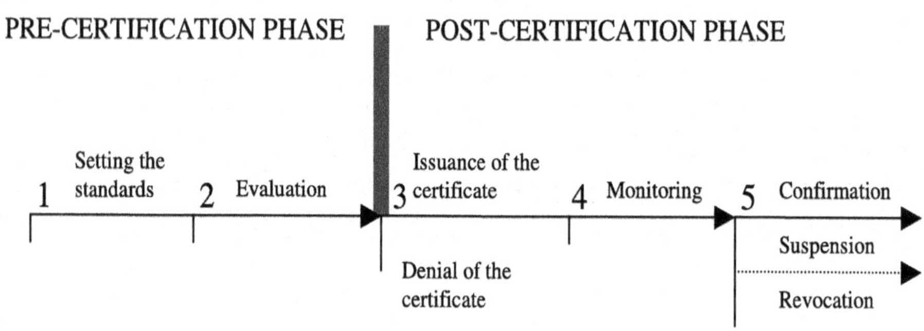

Table 4. Certification process

[135] See Chapter 2, Section 3 *supra*.

[136] On the importance of certification for e-Commerce and the related legal issues, see Stuurman, K. (1990), p. 75. See also extensively on the matter Brun, M.B. (2001).

2.1 Setting the standards

2.1.1 *The role of standards*

It can be argued that standards are yardsticks by which different products and services may be compared. The comparison of products and services is a primary need in the international market. Drafting standards is the first step to fulfil this need. In a certification process, it is particularly important to use uniform standards so that products and services could be evaluated and compared worldwide according to the same rules.[137] There are national and international organisations that set uniform standards in different sectors; sometimes the market itself is free to develop its own standards, which may eventually become recognised worldwide.

2.1.2 *Official and de facto standards*

Standards can be classified on the basis of different criteria. However, for the purpose of this legal analysis, a distinction between official standards and *de facto* standards may suffice.

Official standards (sometimes also referred to as 'formal' standards) are those that have been developed by national and international standardisation institutes and organisations. The International Organisation for Standardisation (ISO) defines (official) standard as follows: a '[d]ocument established by consensus and approved by a recognized body, that provides, for common and repeated use, rules, guidelines or characteristics for activities or their results, aimed at the achievement of the optimum degree of order in a given context.'[138] Official standards are open and public: the participation in their development is open to all, and the results of the standardisation process are publicly available.

A *de facto* standard differs from a formal standard in the sense that it has not been developed by an official body. *De facto* standards can differ sub-

---

[137] 'In England no less than the Magna Carta legislated for one weight and one measure throughout the land, giving early emphasis to the importance of standardization and accountability' Rothery, B. (1996), pp. 3-4.

[138] See <http://isotc.iso.org/livelink/livelink.exe/fetch/2000/2122/687806/Glossary.htm?nodeid=2778927&vernum=0>.

stantially in origin, nature, and status. They can be closed in the sense that only one organisation or a closed group of organisations can define or update them. The latter applies, for instance, to technical specifications that have initially been developed by a supplier to support his particular product range or marketing strategy and, in the course of time, have become more widely used, also by other suppliers seeking a share of the business. A closed *de facto* standard may be public or non-public, depending on whether the technical specifications are available to everybody or only to the parties that participate in the body that originated the standard. In other words, when *de facto* standards are open, more organisations are involved in their definition and update through an open process, and the results (e.g., the technical specifications) are public, i.e., available to everyone who wants to use them.[139]

2.1.3 *Pros and cons of official and* de facto *standards*

Official standards have the advantage of being, by definition, public. They are accessible to all the actors in the market, which is very unlikely to happen with *de facto* standards. However, since official standards imply some sort of public intervention in regulating the market, they can have some undesirable effect on market growth.

De facto standards are created by the companies themselves. The advantage of this form of self-regulation is its flexibility. This type of standards can be easily updated in order to comply with market needs. A big drawback of *de facto* standards is that they are closed. Only one organisation or a closed group of organisations can define or update them.[140]

2.1.4 *Firms, governments, and users all benefit from standardisation*

Standardisation can be seen as the process of setting out international market rules. In this way, standardisation increases international competition.

---

[139] See Brun, M.B. (2001), p. 30.

[140] In fact, most of the time, large companies tend to elaborate their own internal standards, then claim copyright on them, and subsequently sell them or make the standards available through a licence contract. As a consequence of this policy, access to the standards by market competitors turns out to be difficult and limited. This practice is more likely to create a monopoly of standards and create an unfair competitive advantage for the companies which can join the programme in relation to the other competitors. Ultimately, this situation is not in the best interest of the market. See Brun, M.B. (2001) pp. 56-58.

Consequently, it may help firms trying to expand their market share and hurt those trying to defend their market against others. For governments, standardisation of Information Technology (henceforth: IT) products and services is an important instrument to regulate trade and industry. From the user point of view, standardisation of IT products and services may be important in receiving both better product and service information and some guarantees concerning their characteristics.

2.2 Evaluation

The second step of the certification process consists of the evaluation of the product, the service, the practice, or the policy to be certified. Although the ways in which the evaluation is carried out differ, the most common mechanism is the audit. 'Audit' usually refers to an official examination of the business and financial records of a company in order to see that they are true and correct. However, an audit can also be understood more generally as a systematic quality verification procedure. There are different types of auditing procedures – some of them very simple, others very complex. The quality level of the evaluation procedure will have an important impact on the quality level of the certification, its scope, and its legal value.

2.2.1 *Internal audit based on internal standards*

This is an evaluating procedure which is carried out by the company itself. It is the company that sets the standards which the products, services, procedures, or policies have to comply with. Furthermore, the company appoints employees who will be in charge of assessing whether the internal standards are respected and, if it is the case, issuing a sort of quality 'certificate'.

The internal audit does not involve any third party, neither in the setting of the standards nor in the assessment of the company's compliance with them. It is not possible to define this procedure as certification because, as already pointed out in Section 1 – 'certification is a procedure by which a third party gives a written assurance that a product, process or service conforms to specific characteristics' – certification implies the presence of a third party.

The outcome of the internal audit based on internal standards is comparable to a guarantee (but not to a certificate) issued by a company that its procedures, products, and services comply with some internal standards. In this context, Alan Couret, in his book *La Certification*, gives the example of the mark 'Woolmark'.[141]

The value of this guarantee is very limited because the company's statements have not been counterchecked by a third party.

2.2.2 *Internal audit based on third-party standards*

This certification practice consists of the companies' self-assessment based on third-party standards. Usually, an authorised representative of the company which asks for the certificate has to fill out a questionnaire prepared by a third party, sign it, and attest that the statements made on the questionnaire are true and accurate on the date submitted and will remain true and accurate for the term of the agreement with the certifier. Then the certifier evaluates the answers given in the returned questionnaire and decides whether or not to issue the certificate. This is a certification system which can give some warranties of quality, since it is based on third-party standards, a company's internal audit, and a third-party countercheck.

2.2.3 *External audit*

The only evaluation procedure which can give strong warranties of accuracy and reliability is the external audit. The quality of a company's products, services, practices, or policies to be certified is evaluated on the basis of official standards by external professional auditors, who will eventually issue a statement. On the basis of the information provided in the statement, the entity in charge of the certification decides whether or not to issue the certificate.

2.3 **Issuance or denial of the certificate**

The pre-certification phase ends with the issuance or denial of the certificate. Moreover, if at the end of the auditing procedure the certificate is issued, the post-certification procedure begins.

[141] Couret, A. et al. (1995), p. 10.

2.4 Monitoring

After the certificate has been issued, the ongoing compliance of the certified company products, services, practices, or policies with the standards of the certification programme is monitored. Passive and active monitoring can be distinguished, depending on who takes the initiative for the monitoring. In the case of passive monitoring, the certifier starts to monitor a certified company as a result of a complaint received on that company's non-compliance with the certifier's programme. In the case of active monitoring, the certifier periodically (and proactively) checks the quality of company's certified products, services, practices, or policies. The frequency and the level of the investigations depend upon the sector of certification. For example, in the financial sector, the normal practice is to check on the certified companies every 30 to 90 days.

2.5 Confirmation, suspension, or revocation

The monitoring procedure is nothing but a re-evaluation of the company's certified products, services, practices, or policies. If, at the end of the re-evaluation, the company still complies with the standards of the certification programme, the certificate will be confirmed. If the company no longer meets the certification standards, the certificate will be revoked. Sometimes, if the certifier encounters non-compliance, it sets a time frame for the company to bring its products, services, practices, or policies in line with the standards. It goes without saying that, if the company meets the certification requirements at the end of the suspension time, it can keep the certificate; if it does not, the certificate will be revoked. A certificate is also revoked when the contract terminates. By doing regular monitoring and consistently using their power to confirm, suspend, and revoke certificates, certifiers can foster the compliance of the certified companies with the certifiers' programmes in the phase after the certificate has been issued.

3. THE KEY ELEMENTS OF A TRUSTWORTHY CERTIFICATION PRACTICE

The analysis of the certification process suggests five necessary conditions for a trustworthy certification practice:

1. Certifier independency
2. Impartiality in the auditing procedure
3. Active monitoring of the certified company
4. Certifier enforcement power
5. Certifier accountability

3.1 Certifier independency

The third-party certifier must be an unbiased, independent entity. There cannot be a mixing of interests with the certification candidate.[142] Bernard Brun, in his paper *Nature et impacts juridiques de la certification dans le commerce électronique sur Internet*, pointed out the situation in which the certifier carries out its activity for business as a blatant example of conflict of interests.[143] Of course, if the certifier makes a business out of its activity, its independency can be compromised by its eagerness to make money. However, with the exception of public certification, most of the private certifiers (e.g., financial auditing companies) run their activity for lucrative purposes. Brun's view is therefore slightly naïve. It is a matter of fact that, nowadays, certification is a business. Certifier independency should not depend on whether the certifier makes money by issuing certificates. Seemingly more relevant to the matter at hand are questions such as:

What are the certification fees paid for? and

Does the certifier receive financial funding from potential clients?

There may be two completely different answers to the first question: (a) certification fees are paid for the issuance of a certificate; and (b) certification fees are paid for the conduct of audit. In the first case, the certifier is obligated to issue the certificate to the company. In the second, the certifier is only requested to audit the company. It goes without saying that the first situation excludes an (independent) evaluation of the company. As to the

---

[142] See Havighurst, C.C. (1994), p. 2; Astrue, M.J. (1994), p. 75.

[143] '(...) *un organisme privé est susceptible d'avoir des intérêts qui peuvent mettre en cause son impartialité dans le processus de certification. L'exemple le plus flagrant serait la situation où le certificateur exerce son activité dans un but lucratif*'. Brun M.B. (2001), pp. 44-45.

second question, a bias will also be created if the certifier receives consistent sponsorship by potential clients. In fact, the certifier will think twice about refusing issuing those companies a certificate denouncing their noncompliance with the certification programme, and eventually suspending or revoking their certificates.

3.2 Impartiality in the auditing procedure

In the pre-certification phase, the company to be certified must be audited on the basis of official standards by an independent party.[144] In this way, the impartiality of the audit is guaranteed,[145] as opposed to the situation in which the audit of the company is carried out by one of its employees.[146]

3.3 Active monitoring of the certified company

The certificate is by definition already 'outdated' the moment it is issued or, more precisely, after the company has been audited. However, it is from then on that the certificate will be displayed as a quality guarantee. Therefore, it is essential to assure the trustworthiness of the certificate by actively running periodic controls to assess that the certified quality of the companies' products, services, practices, or policies is still up to standards.

3.4 Certifier enforcement power

It is also very essential that a certifier has the power to take appropriate measures if a certified company does not comply with the certification programme. A certifier must have the necessary independence from and avoid conflict of interests with the certified companies so that it could suspend or revoke a certificate (e.g., the certifier will not have the power to suspend or revoke the certificate of a company that finances the certifier).[147]

A certifier that cannot, for example, suspend or revoke a certificate if a certified company has been shown not to comply with the certification

[144] On the importance of independent assessment, see Couret, A. et al. (1995), p. 5; Rae, A. et al. (1995), p. 2.

[145] See *supra* Subsection 2.2.3 *External audit*.

[146] See *supra* Subsection 2.2.2 *Internal audit based on third-party standards*.

[147] See Subsection 3.1 *supra*.

programme is 'comparable to a tiger without teeth and nails'. Moreover, it represents a risk to people who rely on its possibly outdated certificates. It creates unfair competition among companies, allowing companies which do not comply with the certification programme to convey trust by displaying a certificate. Last but not least, it represents a risk to the reputation of all certifiers. If a certificate is found on, for example, a product which does not meet the certified quality, this can cause a deep loss of confidence in the whole certification system.[148]

3.5 Certifier accountability

People perceive a certificate as a sort of guarantee[149] of the information provided through the certificate. In fact, the certificate informs the customer of certain quality aspects of a product or service and thus of its use.[150] Certifier liability towards third parties can, on the one hand, enhance the accuracy of the information provided through the certificate and thus improve the trustworthiness of the certification service. On the other hand, certifier accountability for the provision of inaccurate information towards third parties reassures people who rely on the certificate and possibly fosters trust in the certification system.[151]

4. Benchmarked TMOs

Four TMOs based in the US and five TMOs based in Europe will now be presented and evaluated on the basis of the five necessary conditions for a trustworthy certification practice.[152] The aim of this effort is to draw some conclusions on the soundness and trustworthiness of the actual TMO practice. The selection criteria differ between the American and the European

---

148 'Any player in the market suffers reputation damage when there is a failure of one sort or another', Samuel DiPiazza, global head of PricewaterhouseCoopers LLP, commenting on the loss of confidence in the quality of financial auditing, in an interview with the Financial Times on 21 July 2005, quoted in Parker, A. (2005), p. 11.

149 See Rae, A. et al. (1995), p. 6; Dean, H. & Biswas, A. (2001), pp. 41-57.

150 See Stuurman, K. (1990), p. 51.

151 See Brun M.B. (2001), pp. 51-53.

152 See Section 3 *supra*.

TMOs. The US market is dominated by the four TMOs selected. Therefore, they are a representative sample of the US TMO practice. The European market, on the other hand, consists of many small players, none of which has reached critical mass. Moreover, it is usually not so easy to find much information on their practices. The five TMOs were chosen because they made enough information available on their websites to carry out an accurate analysis.

4.1 American TMOs

TRUSTe, BBBOnLine, WebTrust, and VeriSign are the TMOs which dominate the market in the US.

TRUSTe (Trusted Universal Standards in Electronic Transaction) was set up in the US in 1996 to fill the US government's need for self-regulation in personal data protection.[153] TRUSTe offers its trustmark programmes worldwide. The object of the trustmark programmes is the protection of personal data. The main scopes of TRUSTe's activity are (a) to help consumers gain control over their personal data; (b) to offer companies privacy standards; and (c) to foster self-regulation. TRUSTe has about 2,500 clients.[154]

BBBOnLine is the Internet counterpart of the Better Business Bureau[155] and the trust roles it plays in the physical market. This self-regulatory initiative was launched in 1997 as the second privacy programme, together with TRUSTe, to stress the commitment of the US government towards e-merchant privacy compliance.[156] BBBOnLine's mission is to promote trust and confidence in the Internet through the BBBOnLine Reliability and Privacy Seal Programmes. BBBOnLine's website trustmark programmes allow companies with websites to display the trustmark once they have been evaluated and deemed to meet the programme requirements.[157] To date, about 40,000 e-merchants display BBBOnLine trustmarks.[158]

153 See extensively Farrell, H. (2005).

154 See member list available at <http://www.truste.org/about/member_list.php>.

155 See <http://www.bbb.org/>.

156 For the story of BBBOnLine, see Farrell, H. (forthcoming).

157 See BBBOnLine Programmes available at <http://www.bbbonline.org/business/>.

158 The list of the certified e-merchants is available at <http://www.bbbonline.org/consumer/>.

The WebTrust trustmark programme was created jointly by the American Institute of Certified Public Accountants (AICPA) and the Canadian Institute of Charted Accountants (CICA) and was launched in June 1999. Webtrust provides several trustmarks.

Generally, e-merchants that request WebTrust's trustmarks are audited on the basis of five principles: (1) security, (2) availability (of relevant information for e-consumers), (3) process integrity, (4) online privacy, and (5) confidentiality.[159] The webpage 'Sites with seal' provides a sampling of about 30 e-merchant sites that have been awarded WebTrust's trustmarks.[160]

VeriSign was created in 1995. Although VeriSign presents itself as a TMO which issues security certificates, in fact, it does not. TMOs are defined as independent organisations that offer a system of certification.[161] Certification is 'a procedure by which a third party gives a written assurance that a product, process or service conforms to specific characteristics.'[162] Since VeriSign does not carry out a conformity evaluation procedure, it cannot *sensu stricto* be defined as a TMO. In fact, VeriSign sells security products that come together with a trustmark that purchasers can post on their websites to show that they use VeriSign technology. Because VeriSign is a strong and well-established brand, its trustmarks are perceived by consumers as quality certificates.[163] A very large number of e-merchants display the VeriSign trustmark on their websites.[164]

4.2 **European TMOs**

Confianza Online, Euro-Label, Luxembourg e-Commerce Certified, Thuiswinkel, and Trusted Shops are the chosen European-based TMOs.

Confianza Online is a joint self-regulatory initiative of the Spanish advertising organisation, the *Asociación para la Autorregulación de la Comunicación Comercial* (AUTOCONTROL), and of the Spanish e-com-

[159] The principles are explained in Suitable Trust Services and Criteria, available at <http://www.webtrust.org/download/final-Trust-Services.pdf>.

[160] The list of certified e-merchants is available at <http://www.webtrust.org/abtseals.htm>.

[161] See Chapter 2, Section 5 *supra*.

[162] See Section 1 *supra*.

[163] The literature also considers VeriSign a TMO. See, e.g., Endeshaw, A. (2001), p. 211.

[164] See Existing Customers, available at <http://www.verisign.com/log-in/index.html>.

merce association, the *Asociación Española de Comercio Electrónico* (AECE). It was launched in 2003 to increase consumer confidence in e-commerce by issuing the Confianza Online trustmark to e-merchants that comply with the Confianza Online Ethical Code on e-commerce and interactive advertising. To date, some 200 e-merchants have displayed the Confianza trustmark on their websites.[165]

Euro-Label is a pan-European self-regulatory initiative launched in November 1999 and coordinated by EuroCommerce,[166] a European association representing the retail, wholesale, and international trade sectors. E-merchants may apply for the Euro-Label trustmarks by contacting their national Euro-Label certification body.[167] This organisation will then carry out a standardised certification process, checking the e-merchant's commercial practices against both the European code of conduct[168] and any specific national commercial rules that apply. If the e-merchants are found to operate according to these rules of commercial practice, they will be granted the Euro-Label trustmark. Euro-Label's main objective is to foster the growth of national and cross-border e-transactions within Europe by ensuring that there is a common basis for online trading that is trustworthy and fair. To that end, Euro-Label guarantees e-consumers that, when an e-merchant has been awarded the Euro-label trustmark, they will receive reliable and trustworthy services. So far, slightly more than 500 e-merchants have been awarded the Euro-Label trustmark.[169]

---

[165] See the list of certified e-merchants available at <http://www.confianzaonline.org/quienessomos/adheridas.php>.

[166] See <http://www.eurocommerce.be>.

[167] Currently, there are six Euro-Label certification bodies: 1. Das Österreichische E-Commerce Gütezeichen in Austria; 2. Labelsite in France; 3. Geprüfter Online-Shop in Germany; 4. ConfCommercio in Italy; 5. Confederación española de comercio / Confianza Online in Spain; 6. Institute of Logistics and Warehousing (Instytut Logistyki i Magazynowania) in Poland.

[168] See <http://www.euro-label.com/euro-label/ControllerServlet;jsessionid=7404E097F309E712B247B672E842E218?lg=en&to=welcome&tc=classic2&z2Page=eurococ&z3Page=blank&tradersQueryParam2=1>.

[169] See the list of certified e-merchants available at <http://www.euro-label.com/euro-label/ControllerServlet;jsessionid=7404E097F309E712B247B672E842E218?lg=en&to=welcome&tc=tradersList&tradersQueryParam=1&tradersQueryParam2=1&tradersQueryWord=&tradersQueryCategoryId=&tradersQueryCountryId=&tradersQuerySort>.

Luxembourg e-Commerce Certified was launched in March 2002. It is a joint initiative of the *Ministère de l'Economie et du Commerce extérieur*, the *Chambre de Commerce*, and the *Chambre des Métiers* to guarantee the quality and security of Luxembourg's commercial websites. One e-privacy trustmark and two e-commerce trustmarks, depending on whether or not e-merchants provide online payment services, are offered. The issuance of each of the three trustmarks is subject to an audit of candidate e-merchants by the *Société Nationale de Certification et d'Homologation* (SNCH)[170] accredited by the *Office Luxembourgeois d'Accréditation et de Surveillance* (OLAS).[171] So far, about 10 e-merchants have been certified by Luxembourg e-Commerce Certified.[172]

Thuiswinkel is a trustmark launched in December 2001 by Thuiswinkel.org, a Dutch association representing retailers that offer distance-selling services to consumers. The Thuiswinkel trustmark has received the approval of the *Consumentenbond*,[173] the largest Dutch consumer association. E-merchants receive the Thuiswinkel trustmark if they comply with the Thuiswinkel code of conduct, which sets forth rules on privacy and fair business practice.[174] So far, about 800 e-merchants have displayed the Thuiswinkel trustmark on their websites.

Trusted Shops was launched in January 2000. It is a private, self-regulatory initiative created in cooperation with consumer protection agencies. 'The primary objective was to meet the demands made by the leading politicians for better security in the Internet – and to confirm the consumer that this security is here to stay.'[175] In order to receive the Trusted Shops trustmark, e-merchants have to meet the Trusted Shops security standards.[176] According to the figure provided by Trusted Shops, about 2,500 e-merchants are today operating under such standards.[177]

---

170 See <http://www.snch.lu/>.

171 See <http://www.olas.public.lu/>.

172 See the list of certified e-merchants available at <http://www.e-certification.lu/listsit.htm>.

173 See <http://www.consumentenbond.nl/?ticket=nietlid>

174 See <http://www.thuiswinkel.org/onderdeel/thuiswinkelennl/index.asp?kolomA=3950&kolomB=&navid=1>.

175 Trusted Shops – About us. Available at <http://www.trustedshops.com/en/trustedshops/index.html>.

176 See Obligations, available at <http://www.trustedshops.com/en/shops/obligations_en.html>.

177 See <http://www.trustedshops.com/en/trustedshops/index.html>.

5. Evaluation of TMO practice

At this point, be it somewhat circuitous, a quality assessment will be carried out to determine the quality of service provided by TMOs. The five elements of a trustworthy certification practice[178] will form the yardstick against which the quality of the TMO service will be measured (see Table 5).

1 Certifier independency
2 Impartiality in the auditing procedure
3 Active monitoring of the certificate owner practice
4 Certifier enforcement power
5 Certifier accountability

Table 5. Yardstick for the measurement of TMO services quality

5.1 Assessing TMO independency

TMOs can be considered independent if their funding structure and the composition of their board of directors are neutral. However, the analysis of the benchmarked TMOs' practice reveals that independency is not easy to achieve.

The trustmark market is totally unregulated and is characterised by the opportunistic behaviour of the players. E-merchants willing to receive a trustmark have to pay, in the majority of cases, an 'entrance fee' (i.e., a fee to become a member of the trustmark programme)[179] as opposed to a 'com-

[178] See Section 3 *supra*.

[179] This is the case, for example, for four out of the nine benchmarked TMOs.

pliance audit fee' (i.e., a fee to be audited). There are in fact more TMOs which 'sell' certificates than TMOs which provide a real quality evaluation service, which may culminate in the issuance of a certificate to an e-merchant but which may also end with the refusal of the certificate.[180]

However, TMOs should not be branded as the 'bad guys'. Given the absence of regulation, TMOs simply adapt their practice to market demand. In fact, e-merchants are not interested in the highest possible standards of inspections. Instead, their main interest lies in acquiring a trustmark as easily as possible. As strict inspections lower the probability of successful certification, e-merchants have an incentive to choose TMOs with low evaluation standards.[181] As a result, TMOs will act in the same way assuming a given fee, i.e., to minimise their audit cost. Consequently, they can become dependent on their clients through a special form of setting the fee, known in auditing theory as 'low-balling'.[182] According to this theory, auditors, in order to win a contract, set the fee for the first inspection far below their calculated real costs. As profits tend to be realised only in an ongoing business relationship, the annual returns from subsequent inspections represent a 'quasi-rent', since they depend on customer loyalty.[183] Therefore low-balling makes the inspectors undesirably dependent on their clients.[184]

Moreover, if TMOs have to keep the fee as low as possible, they will necessarily seek sponsorship[185] in order to survive, which in turn will have further negative effects on their independence. Almost all the benchmarked TMOs receive financial sponsorship.[186] It is not unusual for the same e-merchants that sponsor a TMO to join that TMO's trustmark programme. Sometimes, representatives of those e-merchants are also on the TMO's board of directors.[187] This direct financial link between the TMO and the members of its trustmark programme may imply a conflict of interest, nota-

---

[180] This is the case for three out of the nine benchmarked TMOs.

[181] See Pierce, B. & Sweeney, B. (2004), pp. 415-441.

[182] See Calegari, M. et al. (1998), pp. 255-275; De Angelo, L.E. (1981), pp. 113-127.

[183] The generation of revenue through user fees (in addition to sponsorship) is a critical success factor for TMOs. See De Bruin, R. et al. (2005) Final Report, p. 99.

[184] See Makkawi, B. & Schick, A. (2003), pp. 797-822.

[185] '[I]n the [TMO] operational phase, user fees are the most significant type of revenues, followed by sponsorships', De Bruin, R. et al. (2005) Final Report, p. 96.

[186] See De Bruin, R. et al. (2005) Final Report, p. 89.

[187] This is the case, for example, for two of the nine benchmarked TMOs.

bly when sanctions have to be enforced. In fact, TMOs will think twice before denouncing the non-compliance with their trustmark programmes by one of those e-merchants and eventually revoking their trustmarks.

Another necessary condition for the TMO to be independent is that the composition of the TMO's board should be a balanced representation of all interested parties. This will ensure the objectivity of requirements for e-merchants as well as maintain an egalitarian approach. However, majority of the benchmarked TMOs' boards are composed only of representatives of e-merchants.[188]

5.2 Assessing TMO impartiality in the e-merchant auditing procedure

As to evaluation mechanisms, internal audits are used more often than external audits.[189] Moreover, e-merchants' practice and policies are checked against closed *de facto* standards.[190] TMOs set forth their own standards, which apply only to the e-merchants that join their programmes. In other words, e-merchants that apply for trustmarks are usually requested to self-assess their policy or practice against the TMO's standards. This internal procedure is usually followed by a (self-)declaration of conformity, on the basis of which the TMO issues the trustmark to the e-merchant.

The self-assessment procedure can take various forms. The 'check-the-box' practice is the most widespread. Authorised representatives of e-companies which ask for certificates have to fill in a questionnaire prepared by the TMO. Then they usually have to sign and attest that the statements made on the questionnaire are true and accurate as of the date submitted and will remain true and accurate for the term of the agreement with the TMO. The TMO evaluates the answers given in the returned questionnaire and decides whether or not the certificate will be issued.

In theory, this certification system can give some warranties on quality based on third-party standards, a company's internal audit, and a third-party countercheck. However, in practice, the evaluation procedure presents some weaknesses. The TMO's countercheck is only a formality; usually, the ques-

---

188 This is the case, for example, for five of the nine benchmarked TMOs.

189 Five of the nine TMOs analysed base their certification practice on an internal audit.

190 See *supra* Subsection 2.2.1 *Internal audit based on internal standards*.

tionnaire is already returned to the TMO together with the fee for the trustmark. Therefore, the success of the system is based upon the good faith and professionalism of the e-companies' representatives and, to some minor extent, upon the quality of the questionnaire. The reason why this practice is so popular rests in its inexpensiveness. Furthermore, it does not require the physical presence of the individuals taking part in the certification process. The same remarks are also valid when the self-assessment consists only of the e-merchant's declaration of conformity with the TMO's standards. There are quite a few TMOs which, for the issuance of their trustmarks, require only the e-merchant's declaration of conformity with their standards and principles.[191]

Definitely less common in the trustmark business is the external audit of the e-merchant's practices and policies.[192] In this procedure, professional auditors check the e-merchants' policies and practices against the TMO's standards. On the basis of the auditors' report, the TMO eventually decides whether to issue the trustmark to the e-merchant. Moreover, the trustmark is actually a hyperlink on which an e-consumer can click to read the full report the auditors have written on the company.[193]

It goes without saying that the TMOs' practice based on an external audit of an e-merchant's policies and practices offers more warranties that the quality standards represented by the trustmark are actually met by the certified e-merchant, at least at the time the trustmark was issued. From that moment on, compliance has to be checked through monitoring activities.[194] However, the TMOs which require this evaluation practice are by far the less successful. There are at least two reasons for this. First, the trustmark fees are higher than the ones asked by TMOs which base their certification on the e-merchants' self-assessments. Obviously, a third-party audit is more expensive than a self-assessment. Second, it is more difficult for e-merchants to get the trustmark. In one case, a declaration of conformity, not even always followed by a self-assessment procedure, of the e-merchant is enough to receive the trustmark. In the other case, third-party auditors have

[191] Three of the nine benchmarked TMOs require only the e-merchant's declaration of conformity to their standards/principles.

[192] Two of the nine benchmarked TMOs carry out an external audit of the e-merchant's practices and policies before they decide whether or not to grant e-merchants the trustmark.

[193] See *supra* Subsection 2.2.3 *External audit*.

[194] See Subsection 5.3 *infra*.

to assess that the e-merchant's policies and practices are in compliance with the TMO standards.

If unregulated, the trustmark market, driven by the opportunistic behaviour of the players, stimulates worst practice (i.e., only an internal audit) as opposed to best practice (i.e., an external audit). Therefore, negative consequences involving e-consumers who rely on the trustmarks are more likely to follow.

5.3 Assessing whether TMOs actively monitor e-merchant practice

Given the possible conflict of interests for the TMO due to the financial link[195] with its clients (i.e., certified e-merchants), it would be best if the monitoring practice were undertaken by an independent third-party body, external to the TMO, or by a dedicated impartial body, composed of a balanced representation of the stakeholders, i.e., e-merchants and e-consumers, operating within the TMO. However, in most of cases, the monitoring of the certified e-merchants rely neither on the intervention of an independent body external to the TMO nor on an impartial body inside the TMO. Often, there is no separation of powers within the TMO; in fact, one single body is in charge of all the trustmark-related activities.

Moreover, concerning the modality of monitoring, Requirement 7 (Monitoring System) of the European Trustmarks Requirements (henceforth: ETR)[196] set forth that:

> 'Trustmark schemes should regularly monitor the subscriber's compliance with the trustmark requirements. This should include random checks of the subscriber's site including mystery shopping.

---

[195] See 'sponsorship system' and 'low-balling effect', Section 5.1 *supra*.

[196] European Trustmark Requirements (ETR) is a code of conduct drafted by the European Consumers' Organisation (Bureau Européen des Unions de Consommateurs – BEUC) in 2000. The ETR aim to provide a high standard of consumer protection in electronic commerce and encourage the sale of goods and services on the internet. The ETR offer a basis for good online practice. They do not seek to override or replace any mandatory provisions at the European level. They are supplementary to legal obligations and do not affect consumers' statutory rights. Trustmark schemes are encouraged to meet or exceed the ETR. These requirements are aimed at general trustmarks for e-commerce directed towards consumers.

> Trustmark schemes should report on the results of the monitoring and of the non-compliance complaints received to the independent third party.
> Trustmark schemes should encourage feedback from consumers and other interested parties.'

The provision encourages TMOs to put in place a system of active (i.e., 'random checks of the subscriber's site including mystery shopping') and passive (i.e., 'feedback from consumers and other interested parties') monitoring activities. Moreover, TMOs should also report on the monitoring results.

Generally, the benchmarked TMOs meet the first two recommendations of the ETR. Almost all TMOs use (or at least they declare to use) the 'mystery shopping' method and provide a feedback system by which e-consumers and other interested parties can report complaints about the e-merchants' practices. However, due to the absence of reported information on the monitoring results, it is very difficult to assess whether TMOs in fact actively monitor e-merchants and whether TMOs do it in a non-discriminatory way by random checks or by systematic monitoring of all the certified e-merchants.

The regularity of checks may vary considerably from one TMO to another. Some TMOs limit themselves to 'regularly' sample sites, while other TMOs have a timeframe which ranges between 90 days and one year.

5.4 Assessing TMO enforcement power

Both consumers and businesses find it very important that there is an effective enforcement infrastructure, including a complaints procedure and alternative dispute resolution.[197] However, almost none of the benchmarked trustmark schemes have a full-scale enforcement structure in place.

Requirement 8 of the ETR lays down what a full-scale enforcement procedure should look like:[198]

---

[197] De Bruin, R. et al. (2005) Final Report, p. 83.

[198] European Trustmark Requirements (ETR). As already mentioned in footnote 13, the UNICE – BEUC e-Confidence project establishes European trustmark requirements (ETR) aiming to provide a high standard of consumer protection in electronic commerce and encourage the sale of goods and services on the Internet. The ETR offer a basis for good online practice. They do not seek to override or replace any mandatory provisions at European

'Trustmark schemes should have an adequate and meaningful enforcement mechanism and should take the necessary steps to ensure that subscribers comply with the trustmark requirements.
Trustmark schemes should ensure that, when the trustmark requirements are not met, subscribers undertake to amend practices to bring them into line with the trustmark requirements within a short period of time.
A list of dissuasive and proportionate sanctions should be established, which could include information on the media and financial fines.
Available sanctions Sanctions available should include the withdrawal of the trustmark when the subscriber fails to take action to comply with the trustmark requirements or seriously or repeatedly fails to comply with them.
The enforcement process should be transparent.
Decisions as regards sanctions should be disclosed to the independent third party.
Trustmark schemes should make available to the public decisions to withdraw the trustmark.'

All the benchmarked TMOs offer a dispute resolution programme. Sometimes the judging body is independent and external to the TMO, sometimes it is an internal committee, and this brings about the issues related to the TMO's possible conflict of interests.[199] However, TMO enforcement procedures generally lack transparency. TMOs do not generally provide statistics on the removal of trustmarks from e-merchants' websites. In the best scenario, some provide general figures about the number of trustmarks issued or sometimes the number of disputes dealt with. In most cases, TMOs do not mention or publish information on sanctions imposed on e-merchants.[200] The lack of information on enforcement outcomes makes it very hard to judge the effectiveness of TMO enforcement systems. It leaves open the question of whether the trustmark programme requirements are only statements of aspiration or whether they can be perceived as meaningful because TMOs have the resources to and will effectively police merchant compliance.

level. They are supplementary to legal obligations and do not affect consumers' statutory rights. BEUC, the European Consumers' Organisation, is the Brussels-based federation of independent national consumer organisations from all the Member States of the EU and from other European countries. See <http://ec.europa.eu/consumers/cons_int/e-commerce/e-conf_working_doc.pdf>.

[199] See Subsection 5.1 *supra*.

[200] With only one exception among the benchmarked TMOs.

5.5 Assessing TMO accountability

The vast majority of the benchmarked TMOs tend to waive their liability for any damage sustained by e-consumers who rely on the trustmarks by using cautionary language, including disclaimers, in contractual clauses or notices which are displayed somewhere on the TMO's website.[201]

There is nothing wrong when an organisation implements certain measures to downplay the liability risk it may face. However, TMOs aim at enhancing e-consumer trust in the quality of e-merchant practices. Disclaimers appear to be odd instruments here. It is very unlikely that e-consumer trust in e-merchant security, privacy, and business practices will be enhanced if those that certify those practices refuse any kind of liability related to the information provided in the certificates. In most cases, however, e-consumers are not even aware of such disclaimers.[202]

A selection of the most frequently used grounds on which TMOs try to limit or disclaim their liability is listed hereunder.

5.5.1 *The information provided in the trustmark 'is not to be relied upon'*

Sometimes, TMOs base the exclusion of their liability for damage to any person acting in reliance on the information they provide on the fact that the information is meant merely for general guidance purposes and is not to be relied upon. For example, a disclaimer may state: 'The information contained in this site is for general guidance on matters of interest only. No liability can be accepted by XXXX, its directors, or employees for any loss occasioned to any person or entity acting or failing to act as a result of anything contained in or omitted from the content of this website, or our conclusions as stated.'

---

[201] Clauses which tend to limit or exclude the liability of TMOs to e-consumers can be found on the 'Disclaimers' page, in 'Terms and Conditions', in the 'Certification Practice Statement', in the Seal Licence Agreement, or in the 'Relying Party Agreement'.

[202] See Chapter 2, Subsection 5.2 *supra*.

5.5.2 *The information provided in the trustmark is to be used 'at your own risk'*

By such clause, TMOs state that users may use the information provided in the trustmarks at their own risk. For example, a disclaimer may provide: 'You agree that your use of the seal is solely at your own risk.'

5.5.3 *Slight negligence liability is excluded*

In some cases, TMOs only accept liability for wilful intent and gross negligence. For example, a disclaimer may provide: 'Within the framework of the statutory regulations, XXXX shall only be liable for wilful intent and gross negligence liability; slight negligence shall be excluded.'

5.5.4 *TMOs use 'AS IS' warranty*

Occasionally, TMOs use the standard 'AS IS' warranty[203] to specify that they issue no warranty that the information provided in the trustmark is accurate or complete. At the same time, they do not warrant the merchantability, fitness for a particular purpose, satisfaction quality, and non-infringement of the trustmark as a product. In other words, they fully exclude any kind of liability. For example, a disclaimer may provide:

> 'You agree that your use of the seal is solely at your own risk. You agree that all such seals are provided on an 'as is' and 'as available' basis, except as otherwise noted in this seal licence agreement. XXXX expressly disclaims all representations, warranties, guarantees, terms, or conditions of any kind, whether express or implied, including but not limited to the implied warranties of merchantability, fitness for a particular purpose, satisfaction quality and non-infringement to the extent permitted by applicable law. XXXX does not make any representation, warranty of guarantee that the seal will meet your requirements, or that any service will be uninterrupted, timely, secure or error free, nor does XXXX make any representation, warranty or guarantee as to the results that may be obtained from the use of the seal or to the accuracy or reliability of any information obtained through XXXX.'

---

[203] In *Black's Law Dictionary*, 'as is warranty' is defined as: 'A warranty that goods are sold with all existing faults.' Garner, B.A. (2004).

5.5.5 *TMOs invoke an indemnity or hold-harmless clause*

Often TMOs combine the 'AS IS' warranty with the 'indemnity' or 'hold-harmless' clause[204] by which they try to impose an obligation on the e-merchants who received the trustmarks to defend, indemnify, and hold the TMOs harmless from any liability, damages, costs, and expenses which originate from a third-party claim. For example, a disclaimer may provide:

> 'Licensee will defend, indemnify and hold XXXX, and its officers, directors, employees and representatives harmless from and against any liability, damages, costs and expenses, including without limitation reasonable attorneys' fees, in connection with any third party claims against XXXX, its officers, directors, employees or representatives, arising from or relating to the Site, Licensee's use of the XXXX Mark(s) (except for claims that the XXXX Mark(s) or use of the XXXX Mark(s) infringes any trademark, service mark or certification mark rights of third parties) or Licensee's non-compliance with the Privacy Statement(s) (defined in the attached Schedule) or Program Requirements (which are set forth in the attached Schedule); provided that XXXX (i) provides prompt written notice of any such claim, action or demand; (ii) allows Licensee to control the defence and related settlement negotiations, provided, however, that XXXX shall have the right to participate in such defence with counsel of its own choosing at its own expense; (iii) provides Licensee, at Licensee's request, with reasonable assistance in the defence of such claim, action or demand, so long as Licensee reimburses XXXX for XXXX's reasonable out-of-pocket expenses associated therewith; and (iv) Licensee may not settle a claim in a manner that causes XXXX to incur unindemnified liability, take action, or suffer other injury, without XXXX's written consent, which consent shall not unreasonably be withheld.'[205]

---

[204] In *Black's Law Dictionary*, 'indemnity clause' is defined as: 'A contractual provision in which one party agrees to answer for any specific or unspecific liability or harm that the other party might incur. – Also termed hold-harmless clause; save harmless clause.' Garner, B.A. (2004).

[205] Trustmarks are most of the time licensed to e-merchants for a determinate period of time.

6. Conclusions

The hypothesis put forward at the end of the previous chapter, i.e., that TMOs' services seem to be unreliable,[206] has been confirmed by the foregoing analysis of the TMOs' practices.

Unregulated market forces, characterised by the opportunistic behaviour of the market players, are increasingly driving TMOs towards an untrustworthy practice. A low-price policy and the widespread use of private sponsorship put TMO independence under great pressure. Without the necessary independence, the whole TMO certification activity may be compromised. Furthermore, the analysis of the benchmarked TMOs' practices shows that an audit of e-merchants by an independent third party is the exception while e-merchant self-assessment is the rule/ Active monitoring of e-merchant compliance and enforcement procedures are generally in place. However, due to the lack of transparency and information, it is difficult to assess the effectiveness of the present monitoring activities and enforcement practices. It is also legitimate to doubt, first, whether monitoring activities are in fact carried out and, second, whether enforcement procedures are activated at all.

This analysis suggests that TMO practice is rather untrustworthy. The risks for e-consumers to encounter inaccurate trustmarks on e-merchant websites are quite high. For example, TMOs can fail to detect e-merchants' non-compliance with the trustmark programmes, or conflict of interests can prevent TMOs from revoking trustmarks from e-merchants that no longer fulfil the requirements of the trustmark programmes.

The reported situation is very likely to lead to a loss for e-consumers who rely on the trustmarks. Therefore, TMOs try to limit, if not fully exclude, their liability for damage to any person acting in reliance on the trustmark.

At first glance, it seems that e-consumers cannot hold TMOs liable. However:

Are the liability disclaimers valid and will they stand up in court?

---

[206] See Chapter 2, Section 8 *supra*.

Is there enough ground for e-consumers' action against TMOs in tort or in contract?

What is eventually the third-party liability of TMOs in the US and in Europe?[207]

These and other questions will be addressed in the next chapters.

[207] In light of the above, it would certainly be interesting to check possible TMO third-party liability with respect to the 'Unfair Commercial Practices Directive'; and more precisely, pursuant to Arts. 5, 6, and 7 (Directive 2005/29/EC of the European Parliament and of the Council of 11 May 2005 concerning unfair business-to-consumer commercial practice in the internal market and amending Council Directive 84/450/EEC), Directives 97/7/EC, 98/27/EC and 2002/65/EC of the European Parliament and of the Council and Regulation (EC) No 2006/2004 of the European Parliament and of the Council. (*OJ* L 149, 11 June 2005, p. 12). However, this goes beyond the purpose of the present study. Nevertheless, for the ones interested in deepening said liability aspects, see Weatherill, S. & Bernitz, U. (2007); Howells, G. et al. (2006); De Vrey, R.W. (2006).

Chapter 4
UNITED STATES

1. Introduction

The paper *Auditor Liability for Electronic Commerce Transaction Assurance: the CPA/CA WebTrust* written by Carl Pacini and David Sinason is almost the only specific literature found on TMO liability in the US.[208] Therefore, the information published in that paper will be used as the basis for the analysis carried out in this chapter which will then be integrated and updated with the findings of this book. Moreover, to this day, there has been no reported case law which has specifically dealt with TMO third-party liability. However, there are some cases that may affect the third-party liability of professionals for the online provision of inaccurate business or financial information which may apply by analogy to TMO third-party liability.[209]

These cases will be analysed in Section 6. Before that, a bit of background on the issues related to accountant third-party liability in the US will be provided in Section 2. The possible third-party actions against accountants will be listed in Section 3. The doctrinal positions on legal standards to determine third-party accountant liability will be presented in Section 4. Public policy theories will be dealt with in Section 5. In Section 7, TMO third-party liability scenarios will be outlined in case courts do not find relevant cases to apply to TMOs. The conclusions are discussed in Section 8.

---

208 Pacini, C. & Sinason, D. (1999), pp. 479-514.
209 See Pacini, C. & Sinason, D. (1999), p. 484.

2. ACCOUNTANTS – THIRD PARTIES: A CONTROVERSIAL RELATIONSHIP

The history of accountant third-party liability in the US is characterised by a large number of lawsuits. Most of the time, third parties ask auditors for compensation for the damage they suffered relying on lacking or inaccurate information (e.g., accountant failure to detect and report on errors, illegal client acts, and client creditworthiness). There are several reasons for this litigious relationship. Two of them appear very relevant for the present analysis. To put it down very pragmatically: first, auditors and third parties simply seem not to understand each other; second, auditors are generally considered to be more solvent defendants then their clients in a lawsuit.[210]

The first motivation can be traced to the so-called 'expectation gap theory' which 'refers to a difference between auditors' understanding of their function and the expectations of the auditor's role held by investors, creditors, and other users.'[211] In fact, third parties tend to perceive auditors as guarantors of the accuracy of companies' financial statements.[212] Similarly, e-consumers may perceive that a trustmark provides a level of assurance (amounting to a guarantee) regarding transactions conducted over and information placed on the Internet.[213] Many e-consumers may not understand the limitations of the assurances provided in TMOs' engagements (e.g., Terms and Conditions, Certification Practice Statements, or Relying Party Agreements). It is therefore conceivable that e-consumers who lose money in an electronic commerce transaction will seek legal redress from the TMOs who provided assurances through the issuance of trustmarks.

The second reason for the accountants' third-party litigious relationship can be explained by means of the so-called 'insurance hypothesis' according to which 'the auditor is considered a potential indemnifier if an investment or credit loss is experienced.'[214] Accordingly, accountants are

---

[210] See Feinman, J.M. (2003), p. 52; Raspante, J.F. & Rosario, R. (2001), p. 21; Pacini, C. & Sinason, D. (1999), p. 482-484.

[211] Pacini, C. & Sinason, D. (1999), p. 482; see also; Epstein, M.J. & Geiger, M.A. (1994), pp. 60-66; Porter, B. (1993), pp. 49-68; Hanks, J. (1992), pp. 33 et seq.; Costello, J. (1991), p. 267; Chaffee, N. (1988), p. 882.

[212] See Liggio, C. (1975), p. 42.

[213] Houston, R.W. & Taylor, G.K. (1999).

[214] Pacini, C. & Sinason, D. (1999), p. 482. See also Hill, J. et al. (1993), pp. 12-13. Third-party attitude towards auditors does not come out of the blue. In fact, courts have

considered 'deep pockets' as they usually are both well capitalised and often have malpractice insurance, hence they can absorb losses resulting from their own negligence.[215]

The notion of accountants as 'deep pockets' can be extended to assurance service providers such as TMOs.[216] It is not difficult to imagine several scenarios, possibly involving fraud and/or bankruptcy, in which e-consumers have no recourse against a firm which they engaged in electronic commerce. As was already explained, it is very easy and quick to set up a website as well as to shut it down and disappear in the virtual void without leaving any trace.[217] An e-consumer will then have no recourse against, for example, a cunning e-merchant that sets up a commercial website to defraud e-consumers and then disappears.

3. Possible third-party actions against accountants: negligence or negligent misrepresentation

There are two possible claims that a plaintiff in an auditor liability case can bring: first, that the accountant negligently performed the audit; second, that the information the accountant provided to the plaintiff was misleading because the audit was negligently performed.[218] The first claim technically corresponds to an action of negligence whereas the second is to an action of

often adopted the argument that auditors should provide insurance for losses suffered by corporate investors. See, e.g., *Rosenblum* v. *Adler*, 461 A.2d 138 (N.J. 1983); *Spherex, Inc.* v. *Alexander Grant & Co.*, 451 A.2d 1308 (N.H. 1982).

215 See Menon, K. & Williams, D. (1994), p. 327-342; Baber, W. et al. (1995), pp. 386-387; Stice, J.D. (1991), p. 521. Moreover, the US Supreme Court implicitly endorsed the auditor-as-insurer concept in *U.S.* v. *Arthur Young & Co*: 'By certifying the public reports that collectively depict a corporation's financial status, the independent public auditor assumes a public responsibility transcending any employment relationship with the client. The independent public accountant performing this special function owes ultimate allegiance to the corporation's creditors and stockholders, as well as to the investing public. This "public watchdog" function demands that the accountant maintain total independence from the client at all times and requires complete fidelity to the public trust.' 465 U.S. 805, 817-18 (1984).

216 Although it has to be kept in mind that not all TMOs are so well capitalised. See Chapter 3, Sections 4 and 5 *supra*.

217 See Chapter 1, Section 1 *supra*.

218 See Feinman, J.M. (2003), pp. 30-31.

negligent misrepresentation. However, it will be immediately demonstrated that the theoretical differences between the two actions are not reflected in practice as in both cases the general requirements of negligence apply.

In theory, a misrepresentation action requires third-party reliance on the defendant's misrepresentation, pursuant to what is stated in Section 552 (1) of the Restatement Second of Torts. Furthermore, Section 552 (2) limits both the range of third parties and the type of transaction to which the defendant's duty of care is owed. Contrariwise, no specific requirements are set out for a negligent action.[219]

In practice, the differences between misrepresentation action and negligent action are often insignificant. Indeed, the specific requirements of the misrepresentation action are usually applied by the courts in order for the plaintiff to meet the more general requirements of the negligence action. For example, in order to establish the causal link between the breach of duty and the damage, courts require the plaintiff's reliance on the defendant's wrongful statement. Furthermore, the method used to limit the scope of duty established in Section 552 (2) of the Restatement Second of Torts is often used by courts in the duty analysis of negligence.[220]

---

[219] Restatement (Second) of Torts 552 (1977): 'Information Negligently Supplied for the Guidance of Others. (1) One who, in the course of his business, profession or employment, or in any other transaction in which he has a pecuniary interest, supplies false information for the guidance of others in their business transactions, is subject to liability for pecuniary loss caused to them by their justifiable reliance upon the information if he fails to exercise reasonable care or competence in obtaining or communicating the information. (2)[T]he liability stated in subsection (1) is limited to loss suffered (a) by the person or one of a limited group of persons for whose benefit and guidance he intends to supply the information or knows that the recipient intends to supply it; and (b) through reliance upon it in a transaction that he intends the information to influence or knows that the recipient so intends or in a substantially similar transaction.'

[220] See Feinman, J.M. (2003), p. 33. See also *H. Rosemblum Inc.* v. *Adler*, 461 A.2d 138, 146 (N.J. 1983); *Greycas, Inc.* v. *Proud.*, 826 F2d 1560 (7th Circ. 1987); *Barrie* v. *V.P. Exterminators, Inc.*, 625 So. 2d 1007 (La. 1993); *Walpert, Smullian & Blumenthal, P.A.* v. *Katz*, 762 A.2d 582 (Md. 2000). Contrariwise, see *Tri-Professional Realty, Inc.* v. *Hillenburg*, 669 N.E.2d 1064 (Ind. Ct. App. 1996). *In Raritan River Steel Co.* v. *Cherry, Bekaert & Holland*, 367 S.E.2d 609 (N.C. 1988); *Bily* v. *Arthur Young & Co*, 834 P.2d 745 (Cal. 1992).

4. Legal standards to determine liability to third parties for negligence

There are three general approaches on the liability rule: the requirement of near privity, the foreseeability test, and the group and transaction test set forth in Section 552 of the Restatement (Second) of Torts.

4.1 Near privity

The requirement of near privity is actually an evolution of the requirement of privity, according to which a contractual relationship or direct connection between the accountant and the third party should be found in order for the third party to have a cause of action for negligence against the accountant.[221] Through the years, the privity requirement has been eroded, transforming it to the so-called 'near privity'.[222] The first time that near privity was applied to auditor liability was in *Ultramares* v. *Touche*.[223] As J.M. Feinman aptly described in his paper *Liability of Accountants for Negligent Auditing: Doctrine, Policy, and Ideology*,[224] in this case the New York Court of Appeal, considering the threat of indeterminate liability, concluded that liability should be limited to those in privity with the defendant or those for whose benefit the information is provided. However, it was further explained that if the third party is not in a contractual relationship with the defendant accountant, the 'end and aim' of the transaction must have been to provide the audit for the third party's benefit so that, as in *Glanzer* v. *Shepard*, '[t]he bond was so close as to approach that of privity, if not completely one with it.'[225] According to the rule established by the Court's interpretation of Glanzer, 'only if the third party could enforce the defendant's contract as a

[221] See Pennsylvania Supreme Court in *Landell* v. *Lybrand*, 107 A. 783 (Pa. 1919).

[222] See *Glanzer* v. *Shepard*, 135 N.E. 275 (N.Y. 1922). In this case the New York Court found a bin weigher liable for the economic loss suffered by the purchaser who relied on the wrong weight certified by the bin weigher. Notably, the bin weigher was under contract with the seller of beans. However, the court maintained that since the 'end and aim' of the transaction was to provide a service to the purchaser, the latter had an action against the weigher either for negligent performance of its service or as the third-party beneficiary of the weigher's contract with the seller.

[223] 174 N.E. 441 (N.Y. 1931).

[224] Feinman, J.M. (2003), p. 34.

[225] See *supra* n. 223.

third-party beneficiary would it be able to bring the action in negligence.' The near privity rule was then confirmed by the New York Court of Appeal in *Credit Alliance* v. *Arthur Andersen & Co.*[226]

The relevant test to check whether the requirement of near privity is fulfilled in accountant third-party liability cases was actually established by the New York Court of *Appeal in Credit Alliance* v. *Arthur Andersen & Co.* In this case, the Court coined a threefold test according to which the defendant owes a duty to a third party who relies on his representations only if the following conditions are met:

a) 'The accountant must have known that the financial reports were to be used for a particular purpose or purposes;
b) in the furtherance of which a known party or parties was intended to rely; and
c) there must have been some conduct on the part of the accountants linking them to that party, which indicates the accountant's understanding of that party's reliance.'[227]

The 'linking' concept stated in the third condition – i.e., the need for an action carried out by the accountant which links him to the relying party – is defined in *European American Bank* v. *Strauhs & Kay*[228] as a direct communication both in writing and orally and a number of personal meetings between the relying party and the accountant. It is interesting to point out that 'near privity essentially takes accountant liability and other third-party liability cases out of the realm of tort law and places them within the sphere of contract law, allowing recovery only where there is privity or a third-party beneficiary relationship, and under the Credit Alliance linking conduct element, sometimes not even then.'[229]

---

[226] 483 N.E.2d 110 (N.Y. 1985). For an in-depth analysis of the relationship between privity and near privity see Feinman, J.M. (2003), pp. 34-38.

[227] Id. at 118. The Credit Alliance test was later upheld by the New York Court in *Security Pacific Business Credit, Inc.* v. *Peat Marwick main & Co.* 597 N.E.2d 1080 (N.Y. 1992).

[228] 477 N.Y.S.2d 146, (N.Y. App. Div. 1984).

[229] Feinman, J.M. (2003), p. 38.

4.2 Foreseeability

In the mid-1980s, a number of courts departed from the near privity approach to auditor third-party liability cases. In *Rosenblum* v. *Adler*,[230] the leading case, auditor third-party liability cases were dealt with as ordinary liability cases. Negligent misrepresentation was the basis of liability, irrespective of the existence of the near privity requirement between the accountant and the third party, and the rule of liability for foreseeable harm applied.[231] In fact the New Jersey Supreme Court maintained that auditor duty extends to all: (a) whose reliance on the audited statements is reasonably foreseeable by the auditor,[232] and (b) that have been influenced in their decisions by the information provided in auditor statements. Moreover, the court specified further that the statement has to be provided by the audited company for a proper business scope: 'the auditor owes a duty of care to all who obtain a firm's financial statement directly from the audited entity, but owes no such duty of care to those who obtain it from an annual report in a library or from a government file.'[233] *Rosenblum* v. *Adler* was followed straight away by the Wisconsin Supreme Court in *Citizens State Bank* v. *Timm, Schmidt & Co*[234] and a few years later by the Mississippi Supreme Court in *Touche Ross & Co.* v. *Commercial Union Insurance Co.*[235]

Although it was quite popular in the 1980s, since then the foreseeability approach to auditor third-party liability cases has not been followed much by courts.[236]

---

230 461 A.2d 138 (N.J. 1983).

231 See extensively foreseeability, Keeton, W.P. et al. (1984), p. 1001.

232 See Scherl, J.B. (1994), p. 272.

233 Causey, D. (1987), p. 380. Rosenblum radically altered accountants' negligence liability by extending accountants' duty of care to reasonably foreseeable third parties under certain circumstances. See Hagen III, W.W. (1988), p. 189. The holding in Rosenblum is no longer followed even in its own home, as the New Jersey Legislature in 1994 adopted an accountant privity statute that substituted a near privity test for its foreseeability standard. N.J. Stat. Ann. § 2A:53A-25 (West 2003).

234 335 N.W. 2d 361 (Wis. 1983). Like the New Jersey court, the Wisconsin court concluded that accountant liability cases should be decided under ordinary principles of negligence law. According to those principles, 'a tortfeasor is fully liable for all foreseeable consequences of his act except as those consequences are limited by policy factors', id. at 366.

235 514 So. 2d 315 (Miss. 1987).

236 Feinman, J.M. (2003), p. 38.

4.3 Restatement approach

Despite near privity and foreseeability, Section 552 of the Restatement (Second) of Torts[237] is by far the standard which courts use the most in accountants' third-party liability cases.[238] Applied for the first time in the 1968 case *Rusch Factors, Inc.* v. *Levin* by the Federal District Court in Rhode Island, it has been followed by almost all the courts in the US.[239]

As Jay M. Feinman neatly explained in his paper *Liability of Accountants for Negligent Auditing: Doctrine, Policy, and Ideology*, pursuant to Section 552, there are seven conditions which need to be fulfilled in order for a third party to bring an action in negligent misrepresentation against an accountant:

1. the information is false;
2. the accountant supplies the information in the course of his business or in a transaction in which he has a pecuniary interest;
3. the accountant fails to exercise reasonable care in obtaining or communicating the information;
4. the third party justifiably relies on the information, and the reliance causes harm;
5. the third party is the person or is within the group for whom the defendant intends to supply the information or knows that the recipient of the information intends to supply it;
6. the third party relies on the information in a transaction that the defendant intends to influence or knows that the recipient of the information intends to influence, or in a substantially similar transaction; and
7. the third party suffers pecuniary loss.[240]

In most accountant third-party liability cases, the information is inaccurate, it is provided by the accountant in the negligent performance of his profession, and the third party relying on the information suffers pecuniary loss.[241]

---

[237] Restatement (Second) of Torts 552 (1977), see n. 219.

[238] Feinman, J.M. (2003), p. 41.

[239] 284 F. Supp. 85 (D. R.I. 1968). The court expanded accountant liability for negligence from the near privity standard to specifically foreseen or known users. Applying Section 552 of the Restatement (Second) of Torts the Court maintained that an accountant should be liable in negligent misrepresentation for financial misinformation relied upon by actual foreseen and limited classes of persons.

[240] Feinman, J.M. (2003), p. 43.

[241] Therefore, conditions 1, 2, 3 and 7 are either easily met or doctrinally simple.

However,

> *is there enough causal connection between third-party reliance and the loss he suffered?* And
>
> *is third-party reliance justifiable?*[242] Furthermore,
>
> *did the third party intend to supply the information to the third party, or to the group of persons which he belongs to?* Alternatively,
>
> *did the accountant* know *that the recipient of the information intended to supply it to the third party or to the group of persons which he belongs to?*[243] Moreover,
>
> *did the accountant intend to influence the specific transaction, or a substantially similar one, in the course of which the third party relied on the information?* Or instead of that,
>
> *did the accountant* know *that the recipient of the information intended to influence the specific transaction, or a substantially similar one, in the course of which the third party relied on the information?*[244]

These are the key questions which have troubled courts in the application of the Restatement rule.[245]

As to the first two questions on third-party reliance, if the accountant's inaccurate information contributed to the third-party decision, there will be enough causal connection between third-party reliance and the loss he suffered. However, third-party negligent reliance on the accountant's inaccurate information may reduce compensation for the loss suffered or even impede obtaining it from the court.[246] The other questions are particularly significant when the third party is not identified individually in advance

---

[242] These questions relate to condition 4.

[243] These questions relate to condition 5.

[244] These questions relate to condition 6.

[245] See Panttaja, R. (1994), p. 941; *Bethlehem Steel Corp* v. *Ernst & Whinney*, 822 S.W.2d 592, 595 (Tenn. 1991); *Raritan River Steel* v. *Cherry*, 367 S.E.2d 609, 617 (N.C. 1988).

[246] For more information, see *Cordial* v. *Ernst & Young*, 483 S.E.2d 248, at 261-262 (W. Va. 1996); *ESCA Corp.* v. *KPMG Peat Marwick*, 939 P.2d 1228, at 1232 (Wash. Ct. App. 1997).

(e.g., Mr. Jones, the recipient of the accountant's information) but he is a member of a group of persons who may possibly rely on the accountant's information (e.g., Mr. Jones, one of the investors who may possibly rely on the accountant's information). The answers to these questions vary considerably depending on how the courts interpret the concepts of 'to intend' and 'to know'. Courts' interpretation of these two concepts can actually range from: (a) requiring the accountant to have an actual knowledge of the potential reliance without emphasising the accountant's intent and directly taking responsibility, to (b) requiring the accountant to have reason to know of the possible third-party reliance.[247]

5. PUBLIC POLICY

Which of the many available doctrines should the court apply?

How should the court apply the doctrine it has selected?

The answer to these fundamental questions can be given through public policy analysis. Public policy analysis consists of arguments about the so-

[247] See extensively on the matter Feinman, J.M. (2003), pp. 44-48. The leading case for the first type of interpretation is *Bily* v. *Arthur Young & Co.* (834 P.2d 745 [Cal. 1992]) in which the California Supreme Court maintained that: 'The representation must have been made with the intent to induce plaintiff, or a particular class of persons to which plaintiff belongs, to act in reliance upon the representation in a specific transaction, or a specific type of transaction, that defendant intended to influence. Defendant is deemed to have intended to influence [its client's] transaction with plaintiff whenever defendant knows with substantial certainty that plaintiff, or the particular class of persons to which plaintiff belongs, will rely on the representation in the course of the transaction. If others become aware of the representation and act upon it, there is no liability even though defendant should reasonably have foreseen such a possibility.' Id. at pp. 772-773. The leading case for the second type of interpretation is *Blue Bell, Inc.* v. *Peat, Marwick, Mitchell & Co.*, 715 S.W.2d 408, (Tex. App. 1986) in which the Texas Court of Appeals argued that: 'To allow liability to turn on the fortuitous occurrence that the accountant's client specifically mentions a person or class of persons who are to receive the reports, when the accountant may have that same knowledge as a matter of business practice, is too tenuous a distinction for us to adopt as a rule of law. Instead, we hold that if, under current business practices and the circumstances of that case, an accountant preparing audited financial statements knows or should know that such statements will be relied upon by a limited class of persons, the accountant may be liable for injuries to members of that class relying on his certification of the audited reports.' Id. at p. 412.

cial desirability of rules or outcomes in terms of the social values of utility, right, morality, or legal institutional values such as judicial competence and administrability.[248] When doctrine breaks down, and a case or a class of cases cannot be decided within the usual mode of doctrinal analysis, policy analysis prescribes and then interprets the controlling doctrine.[249] Public policy analysis plays a substantial role in court decisions. In Section 6 (where the relevant case law is dealt with), it will be shown that in auditor liability cases, public policy arguments have indeed influenced courts' decisions more than the law's traditional categories of liability.

Generally, no problem arises when an accountant is called upon to prepare an audit report for an identified third party who will eventually rely on it in order to take further decisions. In this case, the accountant owes both his client and the specific third party a duty to prepare the requested information with reasonable care.[250] The accountant knows that the client's intent is to receive the audit for the third party's interest. Moreover, the liability which would arise from the accountant's negligent performance is quite predictable. In all other cases, the courts' agreement on public policy arguments to be applied to accountant third-party liability breaks down. Two different approaches can be distinguished: the contractarian approach (or 'limited liability theory') and the relational approach (or 'expanded liability theory').

5.1 Contractarian approach

Contractarians, in a nutshell, tend to exclude tort law from the realm of accountant third-party liability on the basis of three arguments.[251] First, they see accountant responsibility to clients as more prominent than the one

---

248 See Kennedy, D. (1997), pp. 97-105.

249 'As it is evident in the auditor liability cases, doctrine does not dictate which rule (near privity, negligence, or misrepresentation) or which interpretation of an adopted rule (strict, broad, or in-between readings of the intent and knowledge requirement of section 552, for example) is authoritative.' Feinman, J.M. (2003), p. 49.

250 In providing his professional service, the accountant has an obligation to provide accurate information to both his client and the third party. The accountant knows that the client's intent is to receive the audit for the third party's interest.

251 On the contractarian approach, see, e.g., Epstein, R.A. (1995), pp. 71 90; Huber, P.W. (1988), p. 5; see also Deakin, S. et al. (2003), pp. 224 et seq.

to third parties. Contractarians' starting point is that the accountants' report is not to be considered as a guarantee: (a) on the accuracy of the clients' financial statements; (b) against the clients' fraud in drafting their financial statements; and (c) on the reliability of the clients' financial past in order to predict their future. In fact, the clients prepare their financial statements and thus, in the contractarians' view, the clients assume the primary responsibility for the accuracy of the information herein towards third parties, whereas the accountants' responsibility is secondary.[252] Second, as third parties are actually free to choose whether or not to rely on the information provided by accountants, third parties shall assume the risk accordingly. Third, according to contractarians, by applying tortious liability to accountants' third-party liability cases, courts limit the parties' autonomies to contractually regulate their relationships, disrupt the contractual distribution of rights and obligations, and eventually expose accountants to the risk of indeterminate liability. In fact, this narrow liability approach is most of all functional to avoid the threat of the open-ended liability for accountants.[253] Along this line of thinking, Cardozo in *Ultramares* v. *Touche*[254] maintained that liability for negligence 'may expose accountants to a liability in an indeterminate amount for an indeterminate time to an indeterminate class.' Moreover, contractarians maintain that expanding liability does not create substantial incentives for accountants to carry out their profession with better care because malpractice liability, sanctions by regulators, concern for their reputation, and the threat of professional disciplinary proceedings represent already sufficient incentives.[255]

---

[252] As the California court colourfully summarised in *Bily* v. *Arthur Young & Co.*: 'An auditor is a watchdog, not a bloodhound. As a matter of commercial reality, audits are performed in a client-controlled environment. (…) Moreover, an audit report is not a simple statement of verifiable fact that, like the weight of the load of beans in *Glanzer* v. *Shepard*, can be easily checked against uniform standards of indisputable accuracy. (…) The report is based on the auditor's interpretation and application of hundreds of professional standards, many of which are broadly phrased and readily subject to different constructions. Although ultimately expressed in a shorthand form, the report is the final product of a complex process involving discretion and judgement on the part of the auditor at every stage.' 834 P.2d 745 at 762-763 (Cal. 1992).

[253] See Siliciano, J.A. (1988), p. 1943. For more general information on the issue, Dobbs, D.B. (2000), p. 1282; Atiyah, P.S. (1985), p. 488; Perlman, H.S. (1982), pp. 71-72.

[254] 174 N.E. 441 at 444 (N.Y. 1931).

[255] See Feinman, J.M. (2003), p. 58; Siliciano, J.A. (1988), p. 1953.

5.2 Relational approach

Relationals, in opposition to contractarians, develop their expanded liability theory on extra contractual liability, believing that the role of the law is not limited to the enforcement of the contractual conditions agreed by the parties.[256] Such theory is mainly based on three concepts: accountants' professionalism, independency, and accountants' role as guarantors of the accuracy of companies' financial statements. First, relationals stress the fact that accountants have to be competent and practice their profession with due care. Second, accountants need to be independent. They have to be equidistant from the interests of their clients and of the persons who rely on their accounts.[257] Third, relationals acknowledge that accountants are better positioned than third parties to access and check companies' financial accounts.[258]

The fact that very often accountants' reports are the only source of information on companies' financial situation stimulates an almost complete reliance by third parties on them.[259] Therefore, relationals – unlike contractarians – maintain that the audit provides assurance that things are what they appear to be. As Comptroller General David M. Walker stated after Enron, making special reference to the auditor's role in the financial markets: 'The auditor's opinion on the financial statements is like an expert's stamp of approval to the public and the capital markets.'[260]

On the basis of accountants' professionalism, independency, and their role as guarantors of the accuracy of companies' financial statements, relationals support tortious third-party liability for accountants, thus going beyond liability for pure contractual obligations.[261] Moreover, to the argu-

---

256 On the relational approach, see, e.g., Feinman, J.M. (1995), § 7.3; Macneil, I.R. (1980), pp. 71-117. Feinman, J.M. (2000), p. 737-748; Feinman, J.M. (1990), pp. 1299-1304; Gottlieb, G. (1983), pp. 567-612; Macneil, I.R. (1983), pp. 340-418.

257 See Bailey, L.P. (2002), pp. 16-19.

258 See the concept of information asymmetry dealt with in Chapter 2, Section 4 *supra*.

259 See Winier, H.B. (1983), p. 253.

260 David M. Walker, U.S. Gen. Accounting Office, *Protecting the Public Interest: Selected Governance, Regulatory Oversight, Auditing, Accounting, and Financial Reporting Issues* (GAO-02-483T) 4 (Mar. 5, 2002) (statement before the Committee on Banking, Housing, and Urban Affairs, U.S. Senate).

261 In light of Arthur Andersen's participation in the Enron debacle, supporters of the expanded liability theory consider Cardozo's view in *Ultramares* v. *Touche* outdated. In

ment of potential accountant indeterminate liability, relationals reply that they do not see it as a major issue. They maintain that even if the foreseeability test – the broadest one liability-wise[262] – is applied by courts, accountant liability will be limited to third parties whose reliance on the accountant's report should have been foreseen at the time the report was prepared. In this way, the accountant will be liable to predict his liability and the related potential loss because he should reasonably know the scope of the transactions which the report will be used for.

6. Analysis of the case law potentially applicable to TMOs

As already mentioned at the beginning of this chapter, no reported cases have specifically addressed TMO third-party liability. However, there is some case law on the third-party liability of professionals for the online provision of inaccurate business or financial information that may apply by analogy to TMO third-party liability.

6.1 The Jaillet line of cases

The main doctrine on TMOs[263] has specifically identified a line of cases that originated from *Jaillet* v. *Cashman*[264] (henceforth: the Jaillet line of cases) which may be applied to TMOs. As it will be explained further, the applicability of these cases to TMOs will grant them protection from negligent third-party actions. In other words, it will create a sort of 'safe harbour' for TMOs.[265]

The Jaillet cases consist of the following:

Ultramares v. *Touche*, consistent with the prevailing view of auditing at the time, Cardozo regarded the auditor's report as 'primarily for the benefit of the [auditor's client], a convenient instrumentality for the development of the business, and only incidentally or collaterally for the use of those to whom [the company] might exhibit it thereafter.' 174 NE, 441, 446 (NY 1931). See Allegaert, C. & Tinkelman, D. (2000), pp. 502-503.

262 See Section 4 *supra*.

263 See Pacini, C. & Sinason, D. (1999), p. 484.

264 189 N.Y.S. 743 (Sup. Ct. 1921), aff'd, 194 N.Y.S. 947 (App. Div. 1922), aff'd, 130 N.E. 714 (N.Y. 1923).

265 See Subsections 6.2 and 6.3 *infra*.

1. *Jaillet* v. *Cashman*
2. *Daniel* v. *Dow Jones & Co.*[266]
3. *First Equity Corp.* v. *Standard & Poor's Corp.*[267]
4. *Gutter* v. *Dow Jones, Inc.*[268]
5. *Gale* v. *Value Line*[269]

6.1.1 *Jaillet* v. *Cashman*

As to the facts, in *Jaillet* v. *Cashman*, a New York court held that Dow Jones & Co. was not liable for incorrect information reported on a ticker tape, which caused the plaintiff, who saw the ticker report in his broker's office, to sell certain stocks, resulting in an economic loss. The court held that a provider of financial information was in the same relationship with the public as a newspaper, and there could be no liability for negligence absent a contractual or other special relationship. Furthermore, unless it results from a contract that the provider of information has given an explicit warranty regarding the correctness of the information or a specific 'seal of approval', an action in contract will not be allowed either. Moreover, the court expressed its concern about unlimited liability to an indeterminate class and added very pragmatically that the application of such a rule to the specific case seemed necessary as a matter of practical convenience.[270]

6.1.2 *Daniel* v. *Dow Jones & Co.*

In 1987, a New York trial court faced the issue of the liability of a media company that disseminated incorrect financial data over an electronic network. *Daniel* v. *Dow Jones & Co.*[271] involved Daniel Eldridge, a subscriber to the Dow Jones News Retrieval Service. The service provided instantaneous news transmission by computer-to-computer linkup. Mr. Eldridge brought an action against Dow Jones alleging that he made investment decisions based on false news reports that he received from the Dow Jones

[266] 520 N.Y.S.2d 334 (Civ.Ct. 1987).
[267] 869 F.2d 175 (2d Cir. 1989).
[268] 490 N.E.2d 898 (Ohio 1986).
[269] 640 F. Supp. 967 (D.R.I. 1986).
[270] See Pacini, C. & Sinason, D. (1999), pp. 492-493; Wolfson, J. (1997), p. 83.
[271] 520 N.Y.S.2d 334 (Civ.Ct. 1987).

News/Retrieval service which caused him to lose a substantial sum of money.[272] The plaintiff argued that Dow Jones was liable because the parties had a contract that created a 'special relationship' justifying the imposition of liability for negligent misstatements.

The court, relying on the Jaillet decision, dismissed Mr. Eldridge's claim. In the decision, it was argued that a person who relies on information made electronically available through a computer-to-computer link-up mechanism was comparable to somebody who relies on news provided by paper-based means. From the public policy point of view, the court maintained that the defendant could not face unlimited third-party liability towards an indeterminate class of plaintiffs. In fact, applying the Section 552 of the Restatement (Second) of Tort standard, Mr. Eldridge was not considered 'a person or one of a limited group of persons for whose benefit and guidance' Dow Jones had provided information. Moreover, as the news retrieval service was supplied to more than 200,000 persons, no 'special relationship' was found between the plaintiff and the defendant.[273]

6.1.3 *First Equity Corp.* v. *Standard & Poor's Corp.*

In *First Equity Corp.* v. *Standard & Poor's Corp.*,[274] the plaintiff entered into a one-year subscription agreement with Standard & Poor's (henceforth: S&P) to receive Corporation Records, a loose-leaf summary of corporate finances and operations. First Equity, relying on information provided in Corporate Records, traded convertible secured trust notes issued by Pan America Airways. However, due to the inaccurateness of the information made available by S&P, First Equity suffered losses and consequently brought an action in negligence against S&P. The court, analysing the case at hand, observed that Corporate Records did not make any investment recommendation. Furthermore, a disclaimer placed in the agreement for the provision of Corporation Records set forth that no warranties on the accuracy of the information were made by S&P. The court, relying on the Jaillet rule, maintained that the subscription agreement entered into by and

[272] More precisely, Mr. Eldridge made an investment relying on a price provided on the Dow Jones news report. However, it was not specified that all the prices quoted in the report were expressed in Canadian and not in US dollars.

[273] See Pacini, C. & Sinason, D. (1999), pp. 493-494.

[274] 869 F.2d 175 (2d Cir. 1989).

between First Equity and S&P did not amount to a 'special relationship' between the parties, thus, no extra-contractual duties towards First Equity would arise on S&P. Eventually First Equity's claim against S&P was dismissed.[275]

6.1.4 *Gutter* v. *Dow Jones*

In *Gutter* v. *Dow Jones, Inc.*,[276] a securities investor sued Dow Jones after suffering financial losses by relying on an article published in the Wall Street Journal that incorrectly listed certain bonds as trading with interest. Incidentally, the Ohio Supreme Court had to answer the question of whether a newspaper was accountable to one of its subscribers for a non-defamatory negligent misrepresentation of a fact. On this point, reckoning with the Jaillet rule, the Court held that a subscriber of a newspaper cannot be considered part of a limited class of persons pursuant to Section 552 (2) of the Restatement (Second) of Torts. Hence, absent a 'special relationship', no extra-contractual duty to provide accurate information is owed to a subscriber by a newspaper.

Moreover, the Court backed its purely legal statements by a public policy analysis based on the need to avoid indeterminate liability to an unlimited class of persons and the related chilling effect that the imposition of such liability on news providers may have on their business.[277] Last but not least, the Court added that the defendant fell under the protective effect of the First Amendment, which shields media from liability for negligently providing inaccurate information.[278]

6.1.5 *Gale* v. *Value Line*

In Gale v. Value Line,[279] Stanley Gale entered into a subscription contract with Value Line in order to receive a publication reporting on financial information (i.e., Value Line Convertibles). Due to a publication error, he traded warrants of TransWorld Airlines, relying on incorrect information

[275] See Pacini, C. & Sinason, D. (1999), p. 495.
[276] 490 N.E.2d 898 (Ohio 1986).
[277] See Pacini, C. & Sinason, D. (1999), p. 496.
[278] See, e.g., *Gertz* v. *Robert Welch, Inc.*, 418 U.S. 323 (1974).
[279] 640 F. Supp. 967 (D.R.I. 1986).

published in Value Line Convertibles. Eventually, he had to bear an economic loss. Mr. Gale sued Value Line for negligent misrepresentation citing Section 552 of the Restatement (Second) of Torts. Mr. Gale's negligent misrepresentation claim was barred by the Federal District Court in Rhode Island basically for public policy reasons. As in *Gutter* v. *Dow Jones*, the Court mainly considered the 'chilling' (i.e., adverse) effect that the imposition on publishers of an absolute duty to provide subscribers with correct information may have. In fact, such liability would set an unbearable standard of care for publishers, putting at risk their whole business structure.[280]

6.2 The Jaillet 'safe harbour'

The consistent court decisions in these cases have created a rule which excludes negligent third-party actions against those who negligently disseminate false financial or other information over a public medium – a sort of 'safe harbour'. Actually, the rule goes so far that negligent misrepresentation claims are also restricted to those who have entered into a subscription agreement with the disseminator of business information.[281] However, liability will be imposed where there is a breach of specific contractual obligation regarding information accuracy that might arise from either an express warranty or some indication that the publisher or disseminator has given an express 'seal of approval' to the information.[282] Furthermore, liability will also be imposed where there is breach of fiduciary or other special relationship or duty; or an independent tortious duty, i.e., one imposed by statute or international act such as libel, slander, or fraud.[283] The Jaillet rule does not apply to a specific medium.[284] In fact, the rule has been applied to newspapers,[285] an investment letter,[286] and an interactive computer networks.[287]

[280] See Pacini, C. & Sinason, D. (1999), pp. 496-497.
[281] See *First Equity Corp.* v. *Standard & Poor's Corp.*
[282] Wolfson, J. (1997), p. 83.
[283] See Miller, R. & Young, M. (1997), pp. 1998-1999; Wolfson, J. (1997), pp. 82-83.
[284] See Miller, R. & Young, M. (1997), p. 2000.
[285] See *First Equity Corp.* v. *Standard & Poor's Corp.*; *Gutter* v. *Dow Jones, Inc.*; *Gale* v. *Value Line*.
[286] See *Jaillet* v. *Cashman*.
[287] See *Daniel* v. *Dow Jones & Co.*

The Jaillet rule can be considered a landmark for a number of reasons. In fact, it has been applied:

a) in near privity[288] and in non–near privity jurisdictions;[289]
b) when the court used both near privity[290] and the Restatement (Second) of Torts Section 552 standard;[291]
c) when the third party tried both an action in negligence[292] and in negligent misrepresentation[293] against the disseminator of financial information; and
d) when the parties were bound by a contract[294] and where there was no contractual relationship between them.[295]

As to public policy, the Jaillet line of cases is an example of the consistent application of the contractarian approach (or limited liability theory).[296] In fact, all the standards were interpreted in a restrictive way so that no third-party actions in negligence against one who disseminates false financial or other information over a public medium were allowed. There are mainly two policy considerations on which the courts relied in reaching their decisions:

1. the spectre of unlimited liability to an indeterminate class; and
2. the related 'chilling' effect that may bring the business of information distribution to a standstill, originated by the imposition of a high duty of care on those in this business to a wide class of people.[297]

---

288 See *Jaillet* v. *Cashman*; *Daniel* v. *Dow Jones & Co.*; *First Equity Corp.* v. *Standard & Poor's Corp.*

289 See *Gutter* v. *Dow Jones, Inc.*; *Gale* v. *Value Line.*

290 See *Jaillet* v. *Cashman*; *First Equity Corp.* v. *Standard & Poor's Corp.*

291 See *Daniel* v. *Dow Jones & Co*; *Gutter* v. *Dow Jones, Inc.*; *Gale* v. *Value Line.*

292 See *Jaillet* v. *Cashman*; *First Equity Corp.* v. *Standard & Poor's Corp.*

293 See *Daniel* v. *Dow Jones & Co*; *Gutter* v. *Dow Jones, Inc.*; *Gale* v. *Value Line.*

294 See *Jaillet* v. *Cashman*; *Daniel* v. *Dow Jones & Co.*; *First Equity Corp.* v. *Standard & Poor's Corp*; *Gale* v. *Value Line.*

295 See *Gutter* v. *Dow Jones, Inc.*

296 See Subsection 5.1 *supra.*

297 Pacini, C. & Sinason, D. (1999), pp. 495-496.

6.3 Applicability of the Jaillet rule to TMOs

The prevalent doctrine maintains that the Jaillet rule can be applied to TMOs which will thus be shielded from third-party actions in negligence.[298] In fact, a trustmark and the certificate that comes with the trustmark do not contain any language that constitutes an express warranty or seal of approval,[299] which, following the Jaillet rule, is a necessary condition in order not to be held liable. Rather, TMOs specify in the contractual agreement they use[300] that TMOs do not warrant that the information provided is accurate or complete.

Moreover, TMOs usually state that the information they provide 'is not to be relied upon' and that it can be used only at the recipients' own risk.[301] Therefore, TMOs should not face liability for breach of contract when performing their services because no express warranty or express language exists that will make negligent misstatement a breach of contract. Since near privity would not exist between the TMO and third-party e-consumers, an e-consumer's action against TMOs for negligent misrepresentation of e-merchants' practices would not lie.

As to the other cases, few clarifications are useful to the present analysis. Concerning *Daniel* v. *Dow Jones & Co.*, the court maintained that the First Amendment applied to a computerised news retrieval service.[302] However, this broad interpretation of the scope of the First Amendment may nevertheless not embrace TMOs' services. In fact TMOs' services may not fall within the definition of media services.[303]

In relation to *First Equity Corp.* v. *Standard & Poor's Corp.*, although the case did not involve an accountant, logically the court's reasoning is applicable to accountants, and thus the court's rationale might also be implemented by analogy to TMOs. In fact, S&P's loose-leaf service is comparable to TMO's trustmark service, i.e., to report information on the security,

[298] See Id. pp. 491-500; Miller, R. & Young, M. (1997), pp. 1987-2064.

[299] See Pacini, C. & Sinason, D. (1999), pp. 491-492.

[300] E.g., Terms and Conditions, Certification Practice Statement, Seal Licence Agreement, Relying Party Agreement.

[301] See Chapter 3, Subsection 5.5 *supra*.

[302] The facts in *Daniel* v. *Dow Jones & Co.* may not directly relate to TMO services but the court's opinion recognised the need for the law to adapt itself to changing technology.

[303] See, e.g., Miller, R. & Young, M. (1997), p. 2001.

privacy or business practice of e-merchants with a reasonable degree of accuracy.

6.4 Possible ways to impose negligent liability on TMOs

According to what has been presented so far, TMOs have a good chance of enjoying the 'Jaillet safe harbour' thus protecting them from e-consumers' claims in negligence. However, there may be other ways by which courts impose negligence liability on TMOs.[304]

6.4.1 *Hanberry* v. *Hearst Corp.*

Cem Kaner in his paper *Liability for Defective Content* notably wrote: 'If you promise that your material is safe or accurate, and if it is to your commercial advantage if people rely on your material, and if you invite people to rely on it, then it would be wise to be right. The classic case is *Hanberry* v. *Hearst Corp.*'[305]

As to the facts, in *Hanberry* v. *Hearst Corp.*,[306] an action for negligent misrepresentation was brought against Hearst, a publishing company which had represented in its magazine that specific shoes had 'Good Housekeeping's Consumers' Guaranty Seal' and had certified that the product was a good one. The plaintiff, while wearing the certified shoes, slipped on her kitchen floor, allegedly because the shoes were defective, suffering physical injury as a consequence. The California Court of Appeals recognised the plaintiff's cause of action, considering the defendant's endorsement of the products. Moreover, the Court maintained that while Hearst was not in any contractual relationship with Ms. Hanberry – as with other people — who may have potentially relied on its endorsement, this 'does not mean it is relieved from the responsibility to exercise ordinary care toward them.'[307] Once again, public policy arguments played a major role in the court's decision. The issuance of the 'Good Housekeeping' seal was interpreted by the court as that through which the defendant 'has taken reasonable steps to make an independent examination of the product endorsed.'[308]

304 See Miller, R. & Young, M. (1997), p. 2002.
305 Kaner, C. (1996), p. 56.
306 81 Cal. Rptr. 519 (Ct. App. 1969).
307 81 Cal. Rptr. 519, pp. 684-685 (Ct. App. 1969).
308 Id. at 522. See Pacini, C. & Sinason, D. (1999), pp. 497-499.

6.4.2 *LaSalle National Bank* v. *Duff & Phelps Credit Rating Co.*

In *LaSalle National Bank* v. *Duff & Phelps Credit Rating Co.*,[309] a number of investors, relying on an 'AA' credit rating provided by Duff & Phelps, bought bonds issued by Towers Financial Corporation. The bonds went into default and the investors lodged an action before the New York Court against Duff & Phelps based on negligent misrepresentation. The Court recognised a cause of action on the basis of the Credit Alliance threefold near privity test.[310] In fact, according to the Court: (a) the credit rating was provided to a selected and identified group of investors; (b) the purposes of the investors were known; and (c) the linking requirement was fulfilled by the communications between the defendant and the plaintiffs.[311]

6.5 Checking the potential applicability of Hanberry and LaSalle to TMOs

A number of arguments have been raised by the main doctrine against the applicability of *Hanberry* v. *Hearst Corp.* and *LaSalle National Bank* v. *Duff & Phelps Credit Rating Co.* to TMOs.[312]

As to the Hanberry case, in principle, the court's rationale about the defendant engaging in an 'independent examination' could be applied to an assurance provider, i.e., a TMO. However, *Hanberry* v. *Hearst Corp.* can be distinguished from a typical TMO third-party liability case[313] for a number of reasons.[314] First and foremost, the Hanberry decision is based on Section 311 of the Restatement (Second) of Torts[315] which applies only to physical harm and not to economic losses. Second, it may be argued that

---

309 951 F. Supp. 1071 (S.D.N.Y. 1996).

310 See *supra* Subsection 4.1 *Near privity*.

311 See Pacini, C. & Sinason, D. (1999), p. 499.

312 See Id. pp. 497-498.

313 See Chapter 1, Section 2 *supra*.

314 See Pacini, C. & Sinason, D. (1999), pp. 497-498.

315 Section 311 provides as follows: '(1) One who negligently gives false information to another is subject to liability for physical harm caused by action taken by the other in reasonable reliance upon such information, where such harm results (a) to the other, or (b) to such third person as the actor should expect to be put in peril by the action taken. (2) Such negligence may consist of failure to exercise reasonable care (a) in ascertaining the accuracy of the information, or (b) in the manner in which it is communicated.'

trustmarks do not include any representation as to the quality of the e-merchants' goods or services, or their suitability for any e-consumers' intended purpose. Third, the rules set out in Hanberry v. Hearst Corp. have not been followed much in other cases.[316]

Considering *LaSalle National Bank* v. *Duff & Phelps Credit Rating Co.*, unlike a trustmark/seal of assurance, which may be viewed by anyone with access to the Internet, the Duff & Phelps credit rating was distributed to a select and identifiable group of investors. Moreover, the court found the plaintiffs to have had sufficient communication with Duff & Phelps to satisfy the linking conduct requirement of Credit Alliance.

In a typical e-commerce transaction, however, the e-consumer who purchases goods or services through an e-merchant website will not have any direct communication with the TMO that may be considered sufficient to satisfy the linking conduct requirement. Hence, the facts in LaSalle National Bank are easily distinguishable from an e-commerce transaction made by an e-consumer relying on a trustmark.

6.6 Critical remarks on the possible application of the analysed case law to TMOs

As correctly reported by the prevalent doctrine, when TMOs do not place in their agreements (e.g., Terms and Conditions, Certification Practice Statements, or Relying Party Agreements) or on trustmarks any language that expresses a warranty about the accuracy of the information provided, then TMOs may enjoy the protection granted by the Jaillet 'safe harbour' against third-party action in negligence. However, trustmarks are indisputably displayed on the e-merchant website as seals of approval of e-merchant security, privacy, or business practice. Hence, TMOs may better take seriously

316 See Miller, R. & Young, M. (1997), p. 2004; *Winter* v. *G.P. Putnam's Sons*, 938 F.2d 1033, 1037 (9th Cir. 1991) in which the publisher of The Encyclopaedia of Mushrooms could not be sued for negligent misrepresentation by plaintiffs who became ill after eating poisonous mushrooms because the defendant had not undertaken an independent examination and issued any type of warranty; see also *Yanase* v. *Auto. Club*, 260 Cal. Rptr. 513, 518-19 (Ct. App. 1989) in which the widow and children of a man who was killed by an unknown assailant in a motel parking lot could not sue AAA for negligent misrepresentation because motel accommodations, not neighbourhood safety, were endorsed or represented in an AAA Tourbook.

Cem Kaner's warning '[i]f you promise that your material is safe or accurate, and if it is to your commercial advantage if people rely on your material, and if you invite people to rely on it, then it would be wise to be right.'[317]

In fact, a trustmark gives an unquestionable impression to e-consumers that the TMO has taken reasonable steps to make an independent examination of the certified e-merchant services.[318] The very same impression given by the 'Good Housekeeping' seal made the court in *Hanberry* v. *Hearst Corp.* allow third-party action in negligence against Hearst Corp., the company which issued the seal. Moreover, the argument advanced by the prevalent doctrine that *Hanberry* v. *Hearst Corp.* cannot be applied to a typical TMO third-party liability case because Hanberry is based on a rule which applies only to physical harm and not to economic losses is not conclusive. In fact, courts have already allowed third-party action in negligent misrepresentation, viewing damages for physical injury as analogous to damage for economic loss.[319]

However, a trustmark on an e-merchant website is available and thus can be relied upon by potentially all e-consumers in the world. Therefore, the group of third-party e-consumers tends to be unidentifiable and potentially unlimited. As already pointed out, public policy influences a court's decision more than the law's traditional categories of liability. In fact, if the court will take into account the spectre of unlimited liability and considerations on the chilling effect of imposing a high standard of care on TMOs (involving a wide class of people), future TMO cases will probably enjoy the Jaillet 'safe harbour'.

On the contrary, if the court will regard the trustmark as an 'expert stamp of approval to the public', as Comptroller General David M. Walker urged to consider audit financial statements after the Enron debacle,[320] the applicability of the Jaillet rule will be put under discussion and there may be room for third-party action in negligence against TMOs.

---

[317] Kaner, C. (1996).

[318] See Chapter 2, Section 5 *supra*. See also Dean, H. & Biswas, A. (2001), pp. 41-57; McKnight, H. et al. (2002), pp. 334-359.

[319] See Rosenblum, 461 A.2d at 145-146.

[320] David M. Walker, U.S. Gen. Accounting Office, *Protecting the Public Interest: Selected Governance, Regulatory Oversight, Auditing, Accounting, and Financial Reporting Issues* (GAO-02-483T) 4 (Mar. 5, 2002) (statement before the Committee on Banking, Housing, and Urban Affairs, U.S. Senate).

7. If the case law does not apply to TMOs

What if the US courts apply neither the Jaillet line of cases nor any of the other cited cases to a typical TMO third-party liability case that they will be possibly called to decide?

Which approach to TMO third-party liability for the provision of inaccurate trustmarks will the courts take?

In this section, the application of the three main standards to determine liability to third parties for negligence will be tested on a typical TMO third-party liability case.

7.1 Near privity standard

In order to meet the near privity standard, e-consumers should be 'known parties' to TMOs or, at least, capable of identification. Furthermore, TMOs should be aware of the particular purposes for which e-consumers intend to rely on the trustmarks. On the one hand, TMOs are surely aware that e-consumers will rely on the trustmarks in order to assess e-merchant security, privacy or business practices. On the other hand, the number of e-consumers who may potentially rely on a trustmark is virtually unlimited, hence e-consumers are not capable of identification. On this point, it has been shown in the present analysis that for public policy reasons, courts are reluctant to impose liability on professionals who provide information to an unlimited number of third parties. More precisely, professionals' potential unlimited liability to an indeterminate class and the related 'chilling' effect – originated by the imposition of a high duty of care to a large class of people – which may paralyse their business, are the two main policy concerns.[321]

---

[321] See *supra* Subsections 6.1.1 6.1.3 and 6.1.5 *Jailet* v. *Cashman*; *Daniel* v. *Down Jones & Co.*; *First Equity Corp.* v. *Standard & Poor's Corp*; *Gale* v. *Value Line*. Moreover. See Pacini, C. & Sinason, D. (1999), pp. 495-496; The emblematic Cardozo's statement in *Ultramares* v. *Touche*, where he maintained that liability for negligence 'may expose accountants to a liability in an indeterminate amount for an indeterminate time to an indeterminate class.' 174 N.E. 441 at 444 (N.Y. 1931). See also Siliciano, J.A. (1988), p. 1943. For more general information on the issue, Dobbs, D.B. (2000), p. 1282; Atiyah, P.S. (1985), p. 488; Perlman, H.S. (1982), pp. 71-72.

By the same arguments, courts may not allow a cause of action against TMOs for e-consumers in a typical TMO third-party liability case, not to mention that TMOs will most likely not fulfil the third requirement of the Credit Alliance near privity test: the 'linking requirement'. As regards the case of *European American Bank* v. *Strauhs & Kay*,[322] the 'linking requirement' would mean a direct communication, both orally and in writing, and a series of personal meetings between e-consumers and the TMOs. It is thus reasonable to conclude that under the near privity standard, TMOs would owe no duty to e-consumers who suffer loss by relying on inaccurate trustmarks.

7.2 Reasonable foreseeability

Given a typical TMO third-party liability case, according to the reasonable foreseeability standard, TMOs may owe a duty to provide accurate trustmarks to all e-consumers that TMOs should reasonably foresee to rely upon them and who were in fact influenced in their decision to transact with the e-merchants by the trustmark posted on their websites.[323] However, it has to be stressed that, although the chance that TMOs will be held liable under the reasonable foreseeability standard are quite big, this standard has been followed by very few courts,[324] hence the inherent third-party liability risk to TMOs is very small.

7.3 Restatement

As already pointed out, the 'intent' and 'knowledge' requirements play a crucial role in the restatement test.[325] Courts' interpretation of these two

---

[322] 477 N.Y.S.2d 146, (N.Y. App. Div. 1984). Id.

[323] See Subsection 4.2 *supra*; See also *Rosenblum* v. *Adler*, 461 A.2d 138 (N.J. 1983); Scherl, J.B. (1994), p. 272. Keeton, W.P. et al. (1984), p. 1001; Hagen III, W.W. (1988), p. 189.

[324] See, e.g., Feinman, J.M. (2003), p. 38. Moreover, the holding in Rosenblum is no longer followed even in its own home, as the New Jersey Legislature in 1994 adopted an accountant privity statute that substituted a near privity test for its foreseeability standard. N.J. Stat. Ann. § 2A:53A-25 (West 2003).

[325] See Subsection 4.3 *supra*. Moreover, for more extensive discussion on the matter, see extensively on the matter Feinman, J.M. (2003), pp. 44-48.

concepts can actually range from: (a) requiring the accountant to have an actual knowledge of the potential reliance without emphasising the accountant's intent and directly undertaking of responsibility, to (b) requiring the accountant to have reason to know of the possible third-party reliance.[326] Applying the courts' approach to the typical TMO third-party liability case, if the courts will maintain that TMOs owe a duty to e-consumers (who, to the TMOs' knowledge, are actually aware that they will rely on the trustmarks), the related liability risk to TMOs is almost none given the fact that trustmarks can be relied upon by an unlimited number of e-consumers. However, if courts will allow e-consumers' action when the TMOs have reason to know of their reliance – getting in this way very close to the reasonable foreseeability standard – TMOs may be held liable with respect to aggrieved e-consumers who suffered loss after relying on inaccurate trustmarks.

8. Conclusions

The analysis of the case law, of the body of law potentially applicable to TMOs, of the public policy arguments, and of the doctrine shows that the

[326] The leading case for the first type of interpretation is *Bily* v. *Arthur Young & Co.* (834 P.2d 745 [Cal. 1992]) in which the California Supreme Court maintained that 'the representation must have been made with the intent to induce plaintiff, or a particular class of persons to which plaintiff belongs, to act in reliance upon the representation in a specific transaction, or a specific type of transaction, that defendant intended to influence. Defendant is deemed to have intended to influence [its client's] transaction with plaintiff whenever defendant knows with substantial certainty that plaintiff, or the particular class of persons to which plaintiff belongs, will rely on the representation in the course of the transaction. If others become aware of the representation and act upon it, there is no liability even though defendant should reasonably have foreseen such a possibility.' Id. at pp. 772-773. The leading case for the second type of interpretation is *Blue Bell, Inc.* v. *Peat, Marwick, Mitchell & Co.*, 715 S.W.2d 408, (Tex. App. 1986) in which the Texas Court of Appeals argued that: 'To allow liability to turn on the fortuitous occurrence that the accountant's client specifically mentions a person or class of persons who are to receive the reports, when the accountant may have that same knowledge as a matter of business practice, is too tenuous a distinction for us to adopt as a rule of law. Instead, we hold that if, under current business practices and the circumstances of that case, an accountant preparing audited financial statements knows or should know that such statements will be relied upon by a limited class of persons, the accountant may be liable for injuries to members of that class relying on his certification of the audited reports.' Id. at p. 412.

chances that TMOs will not be liable towards aggrieved third-party e-consumers who have relied on the misleading or false information provided in the trustmark, or in the related certificate, are bigger than the chances that TMOs will be liable.

However, both the case law and the law allow the courts big room for interpretation. In fact, 'as it is in auditor liability cases, the doctrine does not dictate which rule (e.g., privity, near privity, negligence or misrepresentation) or which interpretation of an adopted rule (strict, broad, or in-between readings of, e.g., the "intent" and "knowledge" requirements of Section 552 of the Restatement [Second] of Torts) is authoritative.'[327] Public policy analysis plays a key role in prescribing and then interpreting the controlling rule. The contractarian approach,[328] which supports the limited liability theory, has so far prevailed in public policy analysis of cases concerning professionals who provided inaccurate financial or business information to third parties, also through electronic medium.[329]

The main question in future TMO third-party liability cases is whether the court will decide on the basis of the same public policy considerations. The prevalent doctrine is quite positive about that.[330] However, there are some signals of a possible 'wind of change' in public policy analysis which should not be underestimated. A couple of relevant examples are provided here.

It was recently reported that the FTC considered misleading and unfair the data security practice of a significant number of companies. More precisely, the FTC maintained that such companies misrepresented their data security practices. Furthermore, they did not have reasonable security measures in place.[331] For example, the FTC charged Choice-Point that its security practice violated consumer privacy rights and federal law. Choice-Point agreed to settle the charges. The company had to pay $15 million ($10 million in civil penalties and $5 million in consumer redress). Moreover, the FTC required Choice-Point to implement a comprehensive information

[327] Feinman, J.M. (2003), p. 49.

[328] See Subsection 5.1 *supra*; Moreover, on the contractarian approach, see, e.g., Epstein, R.A. (1995), pp. 71-90; Huber, P.W. (1988), p. 5; see also Deakin, S. et al. (2003), pp. 224 et seq.

[329] See the 'Jaillet line of cases' Subsections 6.1 and 6.2 *supra*.

[330] See, e.g., Pacini, C. & Sinason, D. (1999), pp. 491-501.

[331] See Driver, A.W. (2006), p. 7.

security programme and to obtain audits by an independent third party every two years for twenty years.[332]

CardSystems Solutions was also charged with unfair practices by the FTC. In particular, CardSystems failed to take appropriate security measures to protect sensitive data. It opted for a settlement under which it was required – as with Choice-Point – to implement a comprehensive data security programme and to obtain third-party audits biennially.[333] Hence the FTC seems to assign a significant role to independent third parties which carry out e-merchant security and privacy audits, i.e., TMOs.

Actually, the words that Comptroller General David M. Walker stated after the Enron debacle, making special reference to the auditor's role in the financial markets, may also be quite significant: 'The auditor's opinion on the financial statements is like an expert's stamp of approval to the public and the capital markets.'[334] In fact, the same reasoning can be applied to the role of TMOs in e-commerce, supporting in this way the application of the expanded liability theory (as opposed to the limited liability theory) to TMOs.

However, TMOs have proven not to be very trustworthy.[335]

So *how to stimulate TMOs' best practice?*

John A. Siciliano, in his paper *Negligent Accounting and the Limits of Instrumental Tort Reform*, maintains that expanded liability produces little incentive for accountants to exercise care because they already have sufficient incentives to audit with due care.[336] These incentives include mal-

---

[332] *United States of America (for the Federal Trade Commission)* v. *ChoicePoint Inc. (United States District Court for the Northern District of Georgia, Atlanta Division)*, FTC File No. 052-3069. Available at <http://www.ftc.gov/os/caselist/choicepoint/choicepoint.htm>.

[333] In the Matter of CardSystems Solutions, Inc., and Solidus Networks, Inc., Doing Business as Pay By Touch Solutions, File No. 052 3148. Available at <http://www.ftc.gov/os/caselist/0523148/0523148.htm>.

[334] David M. Walker, U.S. Gen. Accounting Office, *Protecting the Public Interest: Selected Governance, Regulatory Oversight, Auditing, Accounting, and Financial Reporting Issues* (GAO-02-483T) 4 (March 5, 2002) (statement before the Committee on Banking, Housing, and Urban Affairs, U.S. Senate).

[335] See Chapter 2, Section 6 *supra*.

[336] See Siliciano, J.A. (1988), p. 1953.

practice liability to the client, sanctions by regulators in securities cases or other regulated matters, concern for reputation, and the threats of professional disciplinary proceedings.[337] The same argument, however, is not applicable to TMOs. The malpractice liability to client is contractually waved by disclaimers.[338] Furthermore, there are no sanctions by regulators and no threats of professional disciplinary proceedings. Moreover, the only incentive TMOs have is the concern for reputation,[339] which has already proven not to be adequate.[340] Hence, the spectre of third-party liability could be used to bring TMOs' service quality up to the level expected by the FTC.

A counter argument to third-party liability for TMOs is that if TMOs are required to function as insurers of e-commerce transactions – providing a sort of guarantee with related indemnification in case e-consumers experience loss in transactions with e-merchants – e-consumers will enjoy risk-free purchases.[341] Such situation brings about at least two policy concerns. First, having received gratuitously substantial information on e-merchant security, privacy or business practice through the trustmarks, e-consumers may be seen as Internet 'free riders'. Second, it may work as a deterrent for e-consumers to pay due attention in transacting with e-merchants.[342] This argument is in line with the *caveat emptor* doctrine: 'let the e-consumer beware', which is nowadays popular in online consumer protection.[343] However, e-consumers are not able to defend themselves from security and privacy risks because they lack knowledge on these matters.[344] Moreover, *isn't this actually the information gap that TMOs aim to reduce by providing trustmarks?*[345]

---

337 See id.

338 See *supra* Chapter 3 Subsections 5.5.4, 5.5.5 *TMOs use 'AS IS' warranty* and *Indemnity or hold harmless clause.*

339 See Chapter 2 Subsection 5.2 *supra.*

340 See Chapter 2 Section 6 *supra.*

341 See Chapter 2 *supra.* See also Pacini, C. & Sinason, D. (1999), p. 484. More generally, on the same argument applied to third-party accountant liability see Hill, J. et al. (1993), pp. 12-13.

342 See Pacini, C. & Sinason, D. (1999), p. 484; Siliciano, J. A. (1988), p. 1948.

343 'Psst, wanna buy a cheap bracelet?', *The Economist*, 1 July 2004. Available at <http://www.economist.com/research/articlesBySubject/displayStory.cfm?story_ID=2877169&subjectid=1198563>.

344 See the 'information asymmetry issue' dealt with in Chapter 2, Section 4 *supra.*

345 See Chapter 2, Sections 4 and 5 *supra.*

Chapter 5
ENGLAND

1. Introduction

In England, as in the US, no reported legal cases have addressed directly the liability of TMOs. There is also no trace of specific statutory provisions on the matter. Therefore, the analysis of TMO third-party liability will be carried out within the domain of the general principles of professional negligence. More precisely, the relevant legal question that will be answered in this chapter is whether it would be reasonable to impose a duty of care towards e-consumers on TMOs. Looking at the same issue from another perspective, the question can also be presented as whether e-consumers who detrimentally rely on trustmarks and suffer loss will have a cause of action against TMOs that negligently performed their services.

An introduction to third-party duty of care in professional negligence cases will be offered in Section 2. In Section 3, the general rules on the imposition of third-party liability for the provision of negligent misstatement and their applicability to TMOs will be presented. An in-depth analysis of the elements of the duty of care will be carried out in Section 4. In Section 5, a specific line of cases potentially applicable to TMOs will be outlined. The effects of the rules distilled from the line of cases on TMOs' duty of care towards third-party e-consumers will be dealt with in Section 6. In Section 7, the applicability by analogy of CSP third-party liability rules to TMO will be analysed.[346] Section 8 contains the conclusions.

---

[346] See Chapter 1, Subsection 3.1 *supra*.

2. Third-party duty of care in professional negligence cases: a disputed matter

Third-party claims against professionals are usually based on the tort of negligence.[347] In principle, there are three necessary conditions which need to be fulfilled in order to bring an action in negligence:

a) The defendant owes a duty of care to the plaintiff;
b) The defendant has acted or spoken in such a way as to break that duty of care; and
c) The plaintiff has suffered relevant damage as a consequence of the breach.[348]

Therefore, in order to check if a third-party has a cause of action in tort against a professional, the first question to be considered will be whether a professional's duty of care towards the third party exists. Two relevant sub-questions then follow: whether the professional acted with reasonable care in discharging his duty towards the third party, and whether the loss suffered by such third party falls within the scope of the professional duty of care. In England, as in the US,[349] there has been a quite large amount of third-party liability cases brought against professionals in the last 50 years.[350] A couple of reasons for this phenomenon are actually the same as in the US. Third parties tend to see professionals as independent, trustworthy sources of information to rely upon. Furthermore, since professionals are usually

---

[347] See Powell, J.L. & Stewart, R. (2007), pp. 68 et seq.; Lunney, M. & Oliphant, K. (2007), pp. 375 et seq.; Walton, C. et al. (2006), pp. 507 et seq.; Rogers, W.V.H. (2006), pp. 485 et seq.; Dugdale, A.M. & Jones, M.A. (2006), pp. 571-572; Van Dam, C. (2006), pp. 174-179; Hodgson, J. (2004), pp. 123 et seq.; Giliker, P. & Beckwith, S. (2004), pp. 85 et seq.; Deakin, S. et al. (2003), pp. 114 et seq.; Cooke, J. & Oughton, D. (2000), pp. 189 et seq.; Hodgin, R. (1999), pp. 48 et seq.; Zweigert, K. & Kötz, H. (1998), pp. 613-615.

[348] See Powell, J.L. & Stewart, R. (2007), p. 13; See also, Walton, C. et al. (2006), p. 11; Rogers, W.V.H. (2006), pp. 74-75; Dugdale, A.M. & Jones, M.A. (2006), p. 383; Lunney, M. & Oliphant, K. (2007), pp. 120-124; Van Dam, C. (2006), pp. 94-95; Giliker, P. & Beckwith, S. (2004), p. 22; Hodgson, J. (2004), pp. 73-74; Deakin, S. et al. (2003), pp. 74-83; Van Gerven, W. et al. (2000), pp. 44-45; Zweigert, K. & Kötz, H. (1998), p. 609.

[349] See Chapter 4, Section 2 *supra*.

[350] See, e.g., Powell, J.L. & Stewart, R. (2007), pp. 68-69.

insured, they are perceived by third parties as 'deep pockets'.[351] In other words, third parties see professionals as attractive targets to satisfy their request for compensation. Last but not least, the resistance of courts in England to grant damages for negligence causing pure economic loss is lessened in case of negligent information or advice provided by professionals.[352] However, if professionals were liable to all who relied upon them to act with due care, their liability would tend to be unlimited.[353] On the other hand, if professionals were only liable to clients, this would not match the reasonable expectations of society.[354] Courts bear the challenging task of balancing professionals' interests and society expectations.

In order to carry out a preliminary assessment of TMO third-party liability from the very outset, it is significant to present one of the milestones of relevant English case law to understand the method that courts apply in deciding on third-party professional liability: *Candler* v. *Crane, Christmas and Co.*[355] Here the plaintiff, relying on inaccurate (due to negligently prepared) company's accounts, made an investment which eventually lost completely. The Court of Appeal decided that no duty to the plaintiff was owed by the accountants. However, Denning L.J., in his dissenting judgement – later endorsed and consistently relied upon by the House of Lords[356] –

[351] See Powell, J.L. & Stewart, R. (2007), p. 69; Pacini, C. et al. (2002), pp. 427-428. Humphrey, C.G. et al. (1993), pp. 395-411; See extensively, Hartherly, D.J. et al. (1992). See also the literature on the 'expectation gap theory' and the 'insurance hypothesis' quoted in Chapter 4 Section 2 *supra*.

[352] See Powell, J.L. & Stewart, R. (2007), p. 69; Hodgin, R. (1999), pp. 156-157; Witting, C. (2004), pp. 159 et seq. Confront, Lunney, M. & Oliphant, K. (2007), pp. 403 et seq.; Rogers, W.V. H. (2006), pp. 488-489; Hodgson, J. (2004), p. 123.

[353] See Powell, J.L. & Stewart, R. (2007), p. 69; Dugdale, A.M. & Jones, M.A. (2006), p. 571; Witting, C. (2004), p. 320; Giliker, P. & Beckwith, S. (2004), pp. 94-95; Deakin, S. et al. (2003), p. 115. See on the floodgates arguments, Lunney, M. & Oliphant, K. (2007), p. 458; Hodgin, R. (1999), pp. 61-62; Van Dam, C. (2006), p. 170.

[354] See Powell, J.L. & Stewart, R. (2007), p. 69; Witting, C. (2004), pp. 165 et seq.

[355] [1951] 2 K.B. 164.

[356] See *Hedley Byrne & Co. Ltd* v. *Heller and Partners Ltd*, [1964] A.C. 465. See also more recent landmark cases on surveyor' and accountant' liability such as *Smith* v. *Eric S. Bush* [1990] 1 A.C. 844–845; and *Caparo Industries plc* v. *Dickman* [1990] 2 A.C. 621–623 and 656–657. For an in-depth analysis of the influence of the dissenting judgement of Denning L.J. on *Hedley Byrne & Co. Ltd* v. *Heller and Partners Ltd, Smith* v. *Eric S. Bush*, and *Caparo Industries plc* v. *Dickman*, see, e.g., Lunney, M. & Oliphant, K. (2007), pp. 404-429; Giliker, P. & Beckwith, S. (2004), pp. 86-95; Powell, J.L. & Stewart, R. (2007), p. 17; Hodgin, R. (1999), p. 283.

argued in favour of a duty. He basically singled out the three following fundamental questions in order to identify whether a third-party duty of care for the provision of professional statements exists, apart from a contract:

1. 'What persons are under such duty?'[357]
2. 'To whom do these professional people owe this duty?' [358]
3. 'To what transactions does the duty of care extend?'[359]

In the dissenting opinion, Denning L.J. maintained that 'persons such as accountants, surveyors, valuers, and analysts, whose profession and occupation is to examine books, accounts etc. and to make reports on which people – other than their clients – rely in the course of ordinary business' are under a duty of care to third parties.[360] He further explained that the reason why professionals owed the duty was because their calling required particular knowledge and skill. Thus, the level of performance required by the duty is not unreasonable or excessively burdensome. In fact, the special knowledge and skill of professionals work as benchmarks to decide whether or not it is reasonable for a third party to rely on them. Furthermore, the duty is owed to the client and to those third parties with whom the professional deals directly with the purpose of influencing their future conduct (e.g., by making an investment or loan). It also extends to third parties to whom the professional knows his statement will be passed on by his client for the same purpose.[361] This limits the class of those to whom the duty is owed. To translate this concept into legal words, the persons who may possibly count on the information provided by the professional were not confined to clients in contractual relationships, but included all others who met the requirements of proximity.[362] Moreover, Denning L.J. recognised that there are limits to the extent to which it is reasonable to impose a duty of care on a professional as well as limits to the extent to which it is reasonable for a third party to rely on the information provided by the profes-

[357] [1951] 2 K.B. 179.
[358] [1951] 2 K.B. 180-181.
[359] [1951] 2 K.B. 182-183.
[360] [1951] 2 K.B. 179.
[361] See [1951] 2 K.B. 180-181.
[362] See [1951] 2 K.B. 180-181.185.

sional. It is reasonable to impose liability (and for there to be reliance) in case the advice or information is conveyed by the professional himself or by his client in circumstances whether the former is to be taken as knowing that it will be communicated and relied upon. In conclusion, Denning L.J. limited the duty to the specific transaction for which the professional knew that the advice or information was required. In this respect, he stressed the crucial importance of the purposes for which professionals provide information or issue advice.[363]

3. Preliminary consideration on TMO third-party liability and focus on the key question

As anticipated in the previous section, the typical TMO third-party liability case will now be examined in the light of Denning L.J.'s reasoning in order to preliminarily assess TMO third-party liability and identify the crucial issues in order to proceed with the relevant investigations.

Considering the first question answered by Denning L.J.,[364] TMOs can also be seen as professionals under a duty of care towards e-consumers. In fact, as already pointed out in Chapter 1, TMOs can be compared to surveyors, accountants, and, by the same token, to valuers. The TMO profession is to examine e-merchant security, privacy or business practice, and to issue trustmarks (i.e., reports) on which e-consumers (i.e., people other than their clients) rely in the ordinary course of business.

As to the second question,[365] Denning L.J. maintained that accountants' duty (but the same reasoning also applies to surveyors, valuers, and analysts) extends to any person they know that their client is going to show the accounts, with the view of inducing him to invest money or taking some other action on them. However, he added that the duty cannot be extended to strangers of whom accountants have heard nothing and to whom their clients may choose to show their accounts, setting out the proximity test as follows: '*Did the accountants know that the accounts were required for*

---

363 See [1951] 2 K.B. 182–183. On Denning L.J.'s dissenting judgement, see extensively Powell, J.L. & Stewart, R. (2007), pp. 69-72.

364 'What persons are under the duty of care in statement apart from a contract in their behalf?'

365 'To whom do these professional people owe this duty?'

submission to the plaintiff and use by him?'[366] In this way, Denning L.J. recognised that there are limits to the extent to which it is reasonable to impose a duty of care on professionals and thus limits to the extent to which it is reasonable for a third party to rely on the information provided by the professionals. *Mutatis mutandis*, TMOs owe the duty to whom they know e-merchants (i.e., their clients) are going to show the trustmark (which is comparable to accounts[367]) so as to induce the third party to take some action on it. Thus, going down this line of reasoning, TMOs potentially owe the duty to all e-consumers because the trustmark, as evidence of trustworthy security, privacy or business practices, is usually on the homepage of the e-merchant's website to possibly induce all e-consumers to do business online with that e-merchant. In this respect, however, the test of proximity deployed by Denning L.J. does not seem to help limit the number of people to whom the duty is owed. In fact, the TMOs know that the trustmarks were required for submission to possibly all e-consumers and used by them as evidence of trustworthy security, privacy or business practices to rely upon. TMOs owe, then, the duty to a group of people which may be considered definite as a category (e-consumers) but potentially unlimited in number (potentially all e-consumers). Thus, there is the risk of exposing TMOs to a liability to an indeterminate number of people for an indeterminate amount, which, according to the so-called floodgate liability theory, is often not accepted in English law.[368]

Finally, the answer to the third question[369] was given by referring to the very scope of the accounts: 'Only to those transactions for which the accountants knew their accounts were required.'[370] The same can once again

366 [1951] 2 K.B. 180–181.

367 See Chapter 1, Subsection 3.1 *supra*.

368 The rationale of the floodgate liability theory was very well summarised by Cardozo C.J. in *Ultramares Corp.* v. *Touche* who maintained that to allow such recovery would 'expose [defendants] to a liability in an indeterminate amount for an indeterminate time to an indeterminate class', [1931] 255 N.Y. 179. In other words the meaning of this metaphor is that the defendant – who negligently drafts an inaccurate statement which is relied upon by a potentially unlimited number of people and causes them economic losses – would be flooded with claims, resulting in his financial ruin. On the floodgates arguments see extensively, Lunney, M. & Oliphant, K. (2007), p. 458; Hodgin, R. (1999), pp. 61-62; Van Dam, C. (2006), p. 170; Cane, P. (1996), p. 455 et seq.

369 'To what transaction does the duty of care extend?'

370 [1951] 2 K.B. 182-183.

apply to TMOs. Their duty of care extends only to those transactions (to be understood in its broad meaning) for which the trustmark was required. In other words, the duty of care extends to the transactions for which the information in the trustmark is meant. In a nutshell, the duty of care goes as far as the scope of the trustmark goes. However, this is just of little help to limit TMOs' exposure to an open-ended liability for two reasons. First, trustmarks and the information on e-merchant security, privacy or business practices are meant for most of the types of transactions between e-merchants and e-consumers. Second, even if the use of the trustmarks is explicitly limited to only one type of transaction, the number of transactions of that type will remain potentially unlimited.

This preliminary analysis highlights that it is everything but easy to determine whether TMOs owe a duty of care towards e-consumers. In theory, a duty of care can arise by comparing TMOs with accountants, surveyors, and valuers. In practice, however, the imposition of a duty of care on TMOs towards e-consumers gives rise to specific concrete issues, which are mainly related to the possible TMOs' exposure to an unlimited number of third-party claims for an undetermined amount. Consequently, further investigations are needed in order to assess whether it will be possible to impose on TMOs a duty of care towards e-consumers, or, in other words, whether e-consumers who detrimentally rely on trustmarks and suffer loss will have a cause of action against TMOs that negligently performed their services.

Unfortunately, there is no comprehensive test to define the circumstances in which a person owes a duty of care to another, the breach of which causes loss which will give rise to a claim for damages. Instead, the answer is to be found on a case by case basis.[371] To say it with May L.J. in *Merret* v. *Babb*: 'It is reaching for the moon – and not required by authority – to expect to accommodate every circumstance which may arise within a single short abstract formulation. The question in each case is whether the law recognises that there is a duty of care.'[372] Therefore, the most helpful guidance will be obtained by analysing the specific cases in which duties of care have been

---

[371] See Powell, J.L. & Stewart, R. (2007), pp. 525-526; Lunney, M. & Oliphant, K. (2007), pp. 123-124; Dugdale, A.M. & Jones, M.A. (2006), p. 384; Rogers, W.V.H. (2006), p. 134; Witting, C. (2004), pp. 14 et seq.; Giliker, P. & Beckwith, S. (2004), pp. 24-25; Hodgin, R. (1999), p. 63.

[372] [2001] Lloyd's L. Rep. P.N. 468 at para. 41.

imposed. Accordingly, in the next sections, first an analysis of the general principles on duties to third parties will be carried out in order to lay down the basis of the book. Second, the relevant cases in which the duty of care has been imposed will be analysed. After that, there will be enough grounds to provide an informed answer to the key question of whether TMO third-party liability can be found in English law.

4. Duty of care

A duty of care neither arises 'in the air'[373] nor is owed 'to the whole world';[374] generally, a duty – if it arises – is owed to a specific person or to a definable class of persons.[375] This rule applies especially in cases of professional negligent misstatements causing pure economic loss. Foreseeability, proximity, and policy arguments are generally the necessary requirements for the existence of a duty of care.[376] They will be dealt with after having introduced *Hedley Byrne & Co Ltd* v. *Heller & Partners Ltd*,[377] the case which actually opened the way to third-party claims for negligent misstatements causing economic loss and set the standards to verify whether a duty of care exists in such cases.[378] Thus, this is the case to start off with the analysis. As to the facts, in *Hedley Byrne & Co Ltd* v. *Heller & Partners Ltd*, the plaintiffs acted in reliance on favourable references on a company provided by the bankers. However, that company went bankrupt and, as an obvious consequence, the plaintiff suffered economic loss. The House of Lords decided that the bankers did not owe a duty of care towards the plaintiffs because the references were provided 'without responsibility'; in so

[373] *Bourhill* v . *Young* [1943] AC 92, 108, HL (Lord Wright). See also *Donoghue* v. *Stevenson* [1932] AC 562, 618, HL (Lord Macmillan).

[374] *Caparo Industries plc* v. *Dickman* [1990] 2 AC 605, 621, HL (Lord Bridge).

[375] See Witting, C. (2004), p. 14.

[376] Lunney, M. & Oliphant, K. (2007), pp. 124 et seq.; Powell, J.L. & Stewart, R. (2007), pp. 42-49; Dugdale, A.M. & Jones, M.A. (2006), pp. 4391-4397; Rogers, W.V.H. (2006), pp. 158-170; Cooke, J. & Oughton, D. (2000), pp. 159-163.

[377] [1964] A.C. 465.

[378] Lunney, M. & Oliphant, K. (2007), pp. 403 et seq.; Powell, J.L. & Stewart, R. (2007), p. 17; Walton, C. et al. (2006), pp. 516-517; Rogers, W.V.H. (2006), pp. 486-488; Dugdale, A.M. & Jones, M.A. (2006), pp. 444-446; Giliker, P. & Beckwith, S. (2004), pp. 29-30; Cooke, J. & Oughton, D. (2000), pp. 191 et seq.; Hodgin, R. (1999), pp. 51-53.

arguing, the Court implicitly recognised that – contract aside – a professional who issues statements could owe a duty to take reasonable care towards the recipient of the information, breach of which could make the professional liable for third-party economic loss. *Hedley Byrne & Co Ltd* v. *Heller & Partners Ltd* instituted the so-called 'reliance principle' according to which a professional who issues a statement to a person who is entitled to and does rely on it should be liable accordingly. Since then, this principle has been well established. However, quite often courts have used proximity and policy considerations to draw the boundaries of such professional liability.

Therefore, in order to properly understand when a duty of care arises, the analysis will continue by touching upon the straightforward requirement of foreseeability and then focusing on the far more complex proximity and policy arguments. The reader should keep in mind that the three elements are very much interrelated and thus it is not always possible to deal with them in a separate way.[379] Therefore, some flexibility is needed to read the following subsections.

4.1 **Foreseeability of persons**

For the scope of the present analysis, it suffices to say that, as a general rule, the first step that a court takes in order to establish if a duty of care towards the plaintiff may exist for the defendant is to check whether for a reasonable person in the defendant's position it would have been possible to foresee that his carelessness might have caused a loss to the plaintiff, or to the class of persons the plaintiff belongs to.[380] However, in cases of professionals' negligent misstatements causing economic loss to third party who detrimentally relied on them, 'foreseeability' is indeed necessary to establish professional liability but absolutely not sufficient. Such statement was also confirmed by the House of Lords in Caparo Industries plc v. Dickman, in which it was argued as follows:

---

[379] See, e.g., Witting, C. (2004), pp. 25 et seq.

[380] On foreseeability, see extensively, e.g., Dugdale, A.M. & Jones, M.A. (2006), pp. 391-392; Walton, C. et al. (2006), pp. 30-32; Lunney, M. & Oliphant, K. (2007), pp. 129-138; Cooke, J. & Oughton, D. (2000), pp. 159-160.

> 'the postulate of a simple duty to avoid harm that is, with hindsight, reasonably capable of being foreseen becomes untenable without the imposition of some intelligible limits to keep the law of negligence within the bounds of common sense of practicality. Those limits have been found by the requirement of what has been called a 'relationship of proximity' between plaintiff and defendant and by the imposition of a further requirement that the attachment of liability for harm which has occurred be "just and reasonable".'[381]

Hence there is now the need to analyse proximity and then policy arguments.

4.2 **Proximity**

The second step that courts usually take is to assess whether there is enough proximity between the defendant and the plaintiff. 'Proximity' is about the relationship between the parties: being sufficiently proximate, the defendant would know that his failures might directly affect the plaintiff. It plays a significant role in establishing the necessary causal link between the defendant's act and the plaintiff's loss.[382] Unfortunately, there is no simple formula to check in every case whether enough proximity between the defendant and the plaintiff exists.[383] Yet there are many factors which concurrently can help to establish proximity. The most relevant of these factors (i.e., special relationships or relationships equivalent to contracts, mutuality, assumption of responsibility, defendant's skill or special knowledge, defendant's knowledge of the recipient or class, defendant's purpose of information or advice, and plaintiff's reliance) are briefly mentioned here below.[384]

Sometimes courts have asked for proof of the existence of a 'special relationship' between the parties, or of a 'relationship equivalent to con-

---

[381] [1990] 2 AC 633, HL.

[382] See Lunney, M. & Oliphant, K. (2007), pp. 142-143; Dugdale, A.M. & Jones, M.A. (2006), pp. 391-392; Cooke, J. & Oughton, D. (2000), pp. 159-160; Hodgin, R. (1999), pp. 57-58.

[383] In *Caparo Industries plc* v. *Dickman*, the House of Lord maintained that 'there is no simple formula or touch-stone to which recourse can be had in order to provide in every case a ready answer.' [1990] 2 AC 628, HL.

[384] See extensively on the factors influencing proximity Witting, C. (2004), pp. 170-200.

tract'.[385] Other times courts have, nevertheless, pointed out that it does not make much sense to require a relationship equivalent to a contract between the parties of a third-party liability for a negligent misstatement case. In fact, most likely the parties would have never had any kind of communication with each other. However, no matter how the courts call it, what they look for is a relationship which may justify the assumption that the defendant had the knowledge of the loss that his failure could have caused to the plaintiff and the plaintiff trusted in the professional skills of the defendant. Moreover, as it was already stressed, in most of the cases the parties would not have had any contact with each other, but proximity may still be found. If the parties had some relationship, however, this would help very much to establish proximity. In this respect, courts speak about 'mutuality', defined as the fact 'that both plaintiff and defendant played an active part in the transaction from which the liability arose…[T]he plaintiff initiated the relationship by request for a reference; the defendants acted upon the request; and the plaintiffs relied on what they had done.'[386] Mutuality is quite a flexible concept; the proof of it can vary from demonstrating the existence of an ongoing commercial relationship between the parties to a simple plaintiff's request for information or advice to the defendant who might provide it directly or through an intermediary. Needless to say, if the defendant had assumed relevant responsibility toward the plaintiff who suffered loss upon relying on the defendant's negligent statement, the defendant's liability will arise. Furthermore, the concept of 'assumption of responsibility' has been recognised as having a broad meaning. In fact, it can be an express statement of the plaintiff, but also assumed by the facts which consistently point at the recognition of defendant's responsibility (e.g., defendant's undertaking of onerous obligations).[387] A crucial element in order to determine whether it was reasonable for the plaintiff to rely on

385 See, e.g., *Hedley Byrne & Co Ltd* v. *Heller & Partners Ltd* [1964] AC 486, 509, 515, 528, 536, and 539; *Smith* v. *Eric S Bush* [1990] 1 AC 871, HL. See also *Williams* v. *Natural Life Health Foods Ltd* [1998] 1 W.L.R. 835, HL.

386 *White* v. *Jones* [1995] 2AC 283-284, HL.

387 See, e.g., *Hedley Byrne & Co Ltd* v. *Heller & Partners Ltd* [1964] AC 503, HL; *Smith* v. *Eric S Bush* [1990] 1 AC 862, HL; *Williams* v. *Natural Life Health Foods Ltd* [1998] 1 W.L.R. 835, HL. The House of Lords broad interpretation of defendant's 'assumption of responsibility' in these cases reflects the expansion of professional third-party liability for negligent misstatement.

defendant's statement, and thus a necessary element to establish proximity between the parties, is represented by 'defendant's skills or special knowledge'. In order to define such element, it is useful to recall the concept of information asymmetry as explained in Chapter 2, Section 4. Accordingly, there needs to be information asymmetry[388] between the defendant and the plaintiff for proximity to be established. Applying this concept to the situation at hand and simplifying it as much as possible, the plaintiff's reliance on the defendant's statement is justified only when the defendant is a professional on the matter (e.g., he is in the business of providing advice on the relevant issue or has maintained to have the necessary skills or competences to make pertinent statements) and the plaintiff is, on the contrary, ignorant of it or does not have access to the information he needs while the defendant does.[389]

Another proximity factor is the so-called 'defendant's knowledge of the recipient or class'. The overlap between foreseeability and proximity is clear. By 'defendant's knowledge of the recipient or class', what is meant is the defendant's knowledge of the person or the class of persons who are likely to rely on his statement.[390] More precisely, courts do not usually require that the defendant knows the actual identity of the plaintiff, since the reasonable knowledge that the plaintiff, or the class which he belongs to, would have most likely relied on the statement for purposes within the scope of the information would be enough to fulfil this proximity factor.[391] Furthermore, the just mentioned scope for which the information or the advice is provided by the defendant is another proximity factor: the 'defendant's purpose of information or advice'. It assumes crucial importance especially in cases in which the statement issued by the defendant is available to a large number of third parties.[392] However, the defendant's purpose of informa-

[388] Generally, information asymmetry occurs when one party to a transaction has more or better information than the other party. See Chapter 2, Section 4 *supra*.

[389] See, e.g., *Hedley Byrne & Co Ltd* v. *Heller & Partners Ltd* [1964] AC 502-503, 505, 510, and 538-539 HL; *Henderson* v. *Merret Syndicates Ltd* [1995] 2 AC 180 HL.

[390] It is considered a proximity factor because of its function of linking the defendant and the plaintiff. In fact, defendant knowledge of plaintiff reliance brings the latter within the group class of persons directly affected by the act or omission of the former.

[391] See, e.g., *Hedley Byrne & Co Ltd* v. *Heller & Partners Ltd* [1964] AC 465, 482, and 494 HL; *Aiken* v. *Stewart Wrightson Agency* [1995] 1 WLR 1281. Contrariwise, see *Reeman* v. *Department of Transport* [1997] 2 Lloyd's Rep 648, CA.

[392] See Sections 2 and 3 *supra*.

tion or advice does not need to be exactly the one of the plaintiff's reliance on it; it is enough if they are sufficiently congruent.[393] In conclusion, the 'plaintiff's reliance' is an important factor to establish the causal link between the defendant's misstatement and the plaintiff's loss.[394] In fact, a statement causes damage only when somebody acts in reliance on it. However, the plaintiff's reliance needs to be reasonable. It will be considered reasonable if in the case at hand other proximity factors, among the ones already mentioned, coexist (i.e., defendant's specific professional skill and knowledge that the plaintiff or people of the class he belongs to is likely to rely on the statement; and the purposes for which the plaintiff uses the statement are congruent with the one contemplated by the defendant).[395]

4.3 **Policy**

Even if foreseeability and proximity are found, this does not mean that a third-party liability claim for economic loss suffered due to reliance on negligently provided information will go. In other words, foreseeability and proximity are necessary but not sufficient conditions for the recognition of a third-party duty of care upon a professional who negligently provides an inaccurate statement. In addition, the imposition of a duty of care must be fair, just, and reasonable.[396] Fairness, justice, and reasonableness are concepts which enjoy a certain degree of abstraction created to leave some discretion to courts in their decision. Especially in third-party liability claims for negligent misrepresentation causing pure economic loss, which bring about a shared fear for professionals' unlimited liability, these concepts are often used by courts to bar plaintiffs' actions (in jargon: to close the floodgates).[397]

---

[393] See, e.g., *Hedley Byrne & Co Ltd* v. *Heller & Partners Ltd* [1964] AC 503 HL; *Western Trust & Savings Ltd* v. *Strutt & Parker* [1999] PNLR 154, CA. Contrariwise, see *Caparo Industries plc* v. *Dickman* [1990] 2 AC 605, HL where the House of Lords maintained that the identification of the specific purposes for which a statement is issued would not only help in establishing plaintiff's probability of reliance but, more importantly, his entitlement to rely on it.

[394] See Witting, C. (2004), pp. 321-322.

[395] See Id., p. 197.

[396] See, e.g., *Caparo Industries plc* v. *Dickman* [1990] 2 AC 618, HL.

[397] See, e.g., Lunney, M. & Oliphant, K. (2007), pp. 143-144; Dugdale, A.M. & Jones, M.A. (2006), pp. 392-398; Cooke, J. & Oughton, D. (2000), pp. 160-161.

As already pointed out in Section 2, if professionals are under the obligation to act with due care towards all who possibly rely upon the information, the professionals' liability would tend to be unlimited.[398] On the other hand, if professionals are liable only to clients, this would not match the reasonable expectations of society.[399] Courts bear the challenging task to balance professionals' interests and society expectations. Therefore, policy arguments go beyond the pure legal evaluation of the matter at stake and do not only focus on the specific plaintiff and defendant. Courts often consider the potential effects of their decisions on the professional category to which the defendant belongs as a whole, as well as the impacts on parties which may be incidentally affected by the recognition or not of a duty of care.[400] Last but not least, courts may also take into consideration the effect that recognising or not a duty of care may have on their own work and on the administration of justice.

5. Analysis of the case law potentially applicable to TMOs

The study of the elements of third-party liability for negligent misstatement set the theoretical basis for an informed analysis of relevant case law. Presently, no cases of TMOs have been reported in England. However, proceeding with the comparison between TMOs and surveyors and auditors, a line of cases which deals with third-party liability for negligent misstatements under specific conditions (e.g., in the absence of direct dealings between parties, of a defendant's 'assumption of responsibility' for the task, and even of reliance on the plaintiff's statement) may be applied by analogy to the typical TMO cases in which e-consumers suffer a loss from relying on wrong information negligently provided by TMOs through trustmarks.

---

[398] See Powell, J.L. & Stewart, R. (2007), p. 69; Dugdale, A.M. & Jones, M.A. (2006), p. 571; Witting, C. (2004), pp. 165 et seq.; Giliker, P. & Beckwith, S. (2004), pp. 94-95; Deakin, S. et al. (2003), p. 115; See the floodgates arguments, Lunney, M. & Oliphant, K. (2007), p. 458; Hodgin, R. (1999), pp. 61-62; Van Dam, C. (2006), p. 170.

[399] See Powell, J.L. & Stewart, R. (2007), p. 69; Witting, C. (2004), pp. 165 et seq.

[400] See, e.g., *Bolitho* v. *City & Hackney Health Authority* [1998] AC 232, HL; *Phelps* v. *Hillingdon London BC* [2001] 2 AC 619, HL.

5.1 'Extended Hedley Byrne liability' line of cases

As already mentioned at the beginning of the previous section, *Hedley Byrne & Co Ltd* v. *Heller & Partners Ltd* set the ground for tortious liability for the issuance of a negligent misstatement which caused a pure economic loss to third parties that had relied on it. More precisely, for the liability to arise under the *Hedley Byrne & Co Ltd* v. *Heller & Partners Ltd* standard, the circumstances must be 'equivalent to contract' and the defendant must have 'assumed a responsibility' to the plaintiff.[401] However, since *Hedley Byrne & Co Ltd* v. *Heller & Partners Ltd*, the liability for negligent misstatement has evolved. The so-called 'extended Hedley Byrne liability' line of cases has developed.[402] Basically, in such cases, the duty of care arises in the absence of direct dealings between or amongst the parties, of a defendant's 'assumption of responsibility', and even of reliance on the plaintiff's statement. More precisely, as Christian Witting aptly described in his book *Liability for Negligent Misstatements*,[403] the 'extended Hedley Byrne' liability line of cases presents the following specific features:

a) Three parties are involved.
b) It is the third party that claims to have suffered a loss resulting from the inaccuracy of advice or information provided by the defendant.
c) The defendant has no subjective 'assumption of responsibility' for the task.[404]
d) There might be an express exclusion of liability to the plaintiff or to the class he or she belongs to.[405]
e) Ordinarily, there is no mutuality[406] between or amongst the parties. Neither is there a request for advice or information and action by the defendant in direct response to it.[407]

---

401 See Section 4 *supra*.

402 See Lunney, M. & Oliphant, K. (2007), pp. 413 et seq.; Witting, C. (2004), pp. 213 et seq.; Deakin, S. et al. (2003), pp. 124 et seq.; Cooke, J. & Oughton, D. (2000), pp. 199-203.

403 Witting, C. (2004), pp. 213-214.

404 See, e.g., *Smith* v. *Eric S Bush* [1990] 1AC 862, HL; *White* v. *Jones* [1995] 2 AC 256 and 262, HL.

405 See, e.g., *Smith* v. *Eric S Bush* [1990] 1 AC 831, HL.

406 See Section 4 *supra*.

407 See *White* v. *Jones* [1995] 2 AC 271-272, HL.

f) There is often no actual reliance by the plaintiff upon a statement containing inaccurate or misleading advice or information.[408]

On what basis is the duty of care recognised in these cases, anyway?

Smith v. *Eric S Bush* and *Harris* v. *Wyre Forest District Council,*[409] *Spring* v. *Guardian Assurance plc,*[410] and *White* v. *Jones*[411] are the major House of Lords decisions on the 'extended Hedley Byrne liability'.[412] *Smith* v. *Eric S Bush* and *Harris* v. *Wyre Forest District Council* are the most relevant cases for the present analysis, as they present several elements which may be used in the analysis of TMO third-party liability. Hence, they will be dealt with in detail hereunder.

In both *Smith* v. *Eric S Bush* and *Harris* v. *Wyre Forest District Council,*[413] the plaintiffs suffered economic loss from buying real estate from reliance on negligently prepared valuations. In fact, the plaintiffs purchased such real estate through a mortgage loan. Before granting the mortgage, the mortgagees asked the valuers to issue esteems of the relevant real estate, which turned out to be inaccurate. Hence, the plaintiffs purchased the real estates on mortgage loans which were much higher than their actual value. A couple of details crucial to the decisions need to be stressed at this stage: (a) the valuation was at the plaintiffs' expenses (in fact, the plaintiffs paid the valuation fees to the mortgagees); and (b) the real estate properties purchased by the plaintiffs consist of houses at the lower end of the market.

In both cases, the third-party liability disclaimers set forth by the defendants were considered unreasonable and thus void, pursuant to the Unfair Contract Terms Act 1977 (henceforth: UCTA), and the defendants were held liable towards the plaintiffs for their negligent valuations. Literally, Lord Templeman stated that:

> 'in the absence of a (valid) disclaimer of liability the valuer who values a house for the purpose of a mortgage, knowing that the mortgagee will rely

---

[408] See, e.g., *White* v. *Jones* [1995] 2 AC 262 and 272, HL.
[409] [1990] 1 AC 831, HL.
[410] [1995] 2 AC 296, HL. See also *Wade* v. *State of Victoria* [1999] 1 VR 121.
[411] [1995] 2 AC 207, HL.
[412] Witting, C. (2004), pp. 215-221.
[413] [1990] 1 AC 831, HL.

> and the mortgagor will probably rely on the valuation, knowing that the purchaser mortgagor has in effect paid for the valuation, is under a duty to exercise reasonable skill and care and that duty is owed to both parties to the mortgage.'[414]

The House of Lords maintained that the facts showed a high degree of knowledge on the valuers' side that their failure would cause loss to the plaintiffs. The assumption operated by the House of Lords was based on the conviction that purchasers of houses at the lower end of the market almost never commission an independent valuation. They usually rely on the one received by the mortgagees instead. Therefore, the defendants' knowledge of the fact that their failures would most likely directly affect the plaintiffs and cause them loss was considered sufficient to establish proximity between the parties. More precisely, it fulfilled the proximity factor of 'defendant's assumption of responsibility'.[415] The defendants' knowledge of the plaintiffs' probable reliance also played a crucial role in the public policy analysis of the Court, which accordingly established in the defendants a duty of care towards the plaintiffs against the desire of the former that the latter should not rely on the former's valuations. The House of Lords' rationale is well represented by the words of Lord Jauncey, who literally maintained that:

> '[i]t must, however, be remembered that this is a decision in respect of a dwelling house of modest value in which it is widely recognised by surveyors that purchasers are in fact relying on their care and skill. It will obviously be of general application in broadly similar circumstances. But I expressly reserve my position in respect of valuations of quite different types of property for mortgage purposes, such as industrial property, large blocks of flats or very expensive houses.'[416]

---

[414] Id. at 848.

[415] See Section 4 *supra*. On this point, Lord Griffith stated, '[t]he phrase 'assumption of responsibility' can only have any real meaning if it is understood as referring to the circumstances in which the law will deem the maker of the statement to have assumed responsibility to the person who acts upon the advice', [1990] 1 AC 862, HL.

[416] Id. at 859. Such considerations were also relevant to the application of the Unfair Contract Terms Act.

5.2 Duty's factors under the 'extended Hedley Byrne liability' test

A number of recurrent arguments on the necessary elements for the imposition of a duty of care (especially as far as proximity and policy analysis are concerned) may be found in the 'extended Hedley Byrne liability' line of cases.

5.2.1 *Proximity*

From the decisions of the House of Lords, there can be distilled a sort of proximity test used by the Courts in the 'extended Hedley Byrne liability' line of cases. The test is composed of three concurrent requirements.[417]

First, the defendants are professionals in the matter about which they issued the statements, having more knowledge of the issues at stake or better access to relevant information than the plaintiff.[418]

Second, the defendants' knowledge of the fact that their failures will most likely directly affect the plaintiffs and cause them loss are either provable by the plaintiff or clearly indicated by the facts.[419]

Third, the plaintiffs could prove that the defendants knew that the plaintiffs, or the class of persons they belonged to, would likely rely on their statements and thus made an informed decision in undertaking their professional activity. As in the previous requirement, this one could be assumed by the facts.[420]

5.2.2 *Exclusion and limitation of liability clauses*

It is worth pointing out for the present analysis that specifically in both *Smith* v. *Eric S Bush* and *Harris* v. *Wyre Forest District Council*, the valuers disclaimed their responsibilities towards third parties. However, the House of Lords held those disclaimers unreasonable because, pursuant to the UCTA, enough proximity was established between valuers and purchasers. In fact,

---

[417] Witting, C. (2004), pp. 222-225.

[418] See *supra* in Section 4 the proximity factor: 'defendant's skill or special knowledge'.

[419] See *supra* in Section 4 the proximity factor: 'defendant's assumption of responsibility'.

[420] See *supra* in Section 4 the proximity factor: 'defendant's knowledge of the recipient or class'.

the House of Lords rejected the following arguments in which the Council for valuers maintained that it was fair and reasonable for a valuer to rely on an exclusion clause:

a) The exclusion clause was clear, understandable, reiterated, and forcefully drawn to the attention of the purchaser.
b) The purchaser cannot rely on the mortgage valuation and must obtain and pay for his own survey.
c) If valuers cannot disclaim liability, they will face more claims from purchasers, some of whom will be unmeritorious but difficult and expensive to resist.
d) A valuer will become more cautious, take more time, and produce more gloomy reports, which will make house transactions more difficult.
e) If a duty of care cannot be disclaimed, the cost of negligence insurance for valuers, and therefore the cost of valuation fees to the public, will increase.[421]

Lord Templeman replied on these points, as follows:

> 'All these submissions are inconsistent with the ambit and thrust of the act of 1977. The valuer is a professional man who offers his services for reward. He is paid for those services. The valuer knows that 90 per cent of purchasers in fact rely on a mortgage valuation and do not commission their own survey. There is great pressure on a purchaser to rely on the mortgage valuation. Many purchasers cannot afford a second valuation. If a purchaser obtains a second valuation, the sale may go off and then both valuation fees will be wasted. Moreover, he knows that mortgagees, such as building societies and the council...are trustworthy and that they appoint careful and competent valuers and he trusts the professional man so appointed. Finally the valuer knows full well that failure on his part to exercise reasonable skill and care may be disastrous for the purchaser.'[422]

Moreover, it is interesting to emphasise that Lord Griffiths, in supporting Lord Templeman arguments, pointed out a number of additional factors that had influenced the Court's decision on the validity of the valuers' ex-

---

421 See Powell, J.L. & Stewart, R. (2007), p. 553.
422 [1990] 1 AC 852 HL.

clusion of liability clauses: parties' bargaining power, availability of alternative sources of advice (taking into account time and cost), nature of the task undertaken by the defendant (its difficulty or danger), insurability, and practical consequences of (in-) validating the exclusion clause.[423]

5.2.3 *Policy arguments*

As far as policy arguments are concerned, a couple of them are characteristic of the 'extended Hedley Byrne liability' line of cases. In these cases, the House of Lords maintained that it was not only fair, just, and reasonable to place on the defendants a tortious duty of care towards third-party plaintiffs,[424] but also that such duty was not incompatible with the contractual one the professionals owed their clients. Moreover, a defining reason, among the other ones that have already been dealt with,[425] put forward by the House of Lords for the imposition of the third-party duty of care bordered on the fact that the defendants discharged a so-called socially important function.[426] Nevertheless, it has to be kept in mind that generalisations on policy arguments are never recommendable because they depend very much on the specific case. Accordingly, it is useful for the present analysis to point out a crucial policy consideration in the decision to place a duty of care on the valuers in *Smith* v. *Eric S Bush* and *Harris* v. *Wyre Forest District Council*. In these cases, consideration of the nature of the interests in question and the value of the assets purchased played a significant role in the House of Lords' decision on whether it was reasonable to expect the plaintiffs to obtain independent valuations.[427] More precisely, the unreasonableness of requiring the purchasers of real estates at the lower level of the market to pay for a second valuation after having already paid valuation

---

[423] Id. at 858-859 HL.

[424] See Subsection 4.3 *supra*.

[425] See *supra* in this Section *Smith* v. *Eric S Bush and Harris* v. *Wyre Forest District Council.*

[426] That is, a valuer who prepares a valuation of a house at the lower level of the market for the purchasers in *Smith* v. *Eric S Bush* and *Harris* v. *Wire Forest District Council*; an employer who prepares a reference about a former employee who is applying for a new job in *Spring* v. *Guardian Assurance plc.*; a solicitor who carries out instructions for the alteration of a will prior to the death of the testator in *White* v. *Jones.*

[427] See Powell, J.L. & Stewart, R. (2007), p. 553.

fees to the mortgagees was crucial in affirming the imposition of a duty on the valuers in *Smith* v. *Eric S Bush* and *Harris* v. *Wyre Forest District Council*.

6. Applicability of the 'extended Hadley Byrne liability' rules to TMOs

The typical TMO third-party liability case[428] in which an e-consumer suffers loss by relying on an inaccurate trustmark negligently issued by a TMO presents the specific features of the 'extended Hedley Byrne liability' line of cases. In fact:

a) it involves three parties, i.e., the TMO, the e-merchant, and the e-consumer;
b) it is the third-party e-consumer who claims to have suffered loss as a result of the inaccuracy of the information provided by the defendant TMO;
c) there is no assumption of responsibility by the defendant TMO for the issuance of accurate information on e-merchant security, privacy, or business practices provided in the trustmark;
d) (on the contrary,) there is an express exclusion of liability to third-party e-consumers through disclaimers;[429]
e) there is no mutuality of relations between parties. In fact, the TMO does not act in response to an e-consumer's request for information or advice; and
f) the plaintiff e-consumer's actual reliance on the inaccurate information in the trustmark is often absent.

Moreover, the typical TMO third-party liability case presents many similarities with *Smith* v. *Eric S Bush* and *Harris* v. *Wyre Forest District Council*. First, e-consumers may be compared with purchasers of real estates at the lower end of the market. It is very unlikely for such purchasers to obtain a second valuation of houses as it is for e-consumers to obtain a second

428 See Chapter 1, Section 2 *supra*.
429 See Chapter 3, Subsection 5.5 *supra*.

assessment of e-merchant security, privacy or business practices. The reason is that both purchasers of real estates at the lower level of the market and e-consumers do not transact for business purposes. Furthermore, they do not have professional competencies in the transaction at stake. In addition, they usually do not plan to spend more money than the amount necessary to complete the transaction. Often, they cannot even afford a second opinion. Therefore, purchasers of real estates at the lower end of the market are very dependent on valuers as e-consumers are on TMOs. Accordingly, the facts indicate a high degree of knowledge on the part of the defendant TMOs of the probability of loss to the plaintiffs should the auditing of the e-merchant security, privacy or business practices be negligently performed. The only significant difference is that e-consumers do not directly pay for the TMO's service whilst purchasers of real estates at the lower level of the market usually pay valuation fees to the mortgagees. However, it could be argued that e-consumers indirectly pay for the TMO's service. In fact, e-merchants' costs of obtaining trustmarks may be included in the prices of the products or services they offer.

6.1 **Proximity**

Proximity, in a nutshell, is about how one party is placed with regard to another.[430] In this respect, defendant TMOs are professionals that possess a specific knowledge of online security, privacy, or business practices. Furthermore, they are better placed to provide relevant advice than the plaintiff e-consumers, as TMOs have the capability to access more relevant information. The relationship between TMOs and e-consumers is therefore unequal in this way.[431] Moreover, according to the rules on third-party liability for negligent misstatements, the plaintiff e-consumers should provide a number of evidence in order to establish proximity.

First, either it has to be proven or the facts must clearly demonstrate that defendant TMOs had actual knowledge of the likelihood of harming e-consumers should the trustmark service be inadequately performed and, thus,

---

[430] See Lunney, M. & Oliphant, K. (2007), pp. 142-143; Dugdale, A.M. & Jones, M.A. (2006), pp. 391-392; Cooke, J. & Oughton, D. (2000), pp. 159-160; Hodgin, R. (1999), pp. 57-58.

[431] See the concept of 'information asymmetry' in Chapter 2, Section 4 *supra*.

the information related to the e-merchant was inaccurate. As mentioned above, commenting on the similarities between or amongst a typical TMO third-party liability case, *Smith* v. *Eric S Bush*, and *Harris* v. *Wyre Forest District Council*, the facts indicate a high degree of knowledge on the part of the defendant TMOs of the probability of loss to the plaintiffs should the auditing of the e-merchant security, privacy or business practices be negligently performed. However, there is no evidence that 90 percent of e-consumers rely on trustmarks, as it was maintained by Lord Templeman for purchasers of real estates at the lower level of the market in *Smith* v. *Eric S Bush* and *Harris* v. *Wyre Forest District Council.*[432] It could be argued, though, that the TMOs' business model is based on e-consumers' reliance on trustmarks. In fact, if e-consumers do not rely on trustmarks, e-merchants will not demand them; thus, there will be no reason for TMOs to exist. This observation may reinforce the assumption of TMOs' actual knowledge of the likelihood of loss to the plaintiffs if the certification service is inadequately performed.[433]

Second, plaintiff e-consumers have to prove, or it must be clearly indicated by the facts, that TMOs, in light of the aforementioned knowledge, made a conscious decision to provide information on e-merchant security, privacy, or business practices by supplying trustmarks. As the provision of trustmarks represents the core business of TMOs, it should be possible to assume (or not too difficult to prove, anyway) that TMOs, aware of the related risks, consciously decided to enter the business.

Third, plaintiffs would be generally asked to prove the tightness of the causal connection between the TMOs' inadequate performance of the service and the subsequent damage they suffered. E-consumers would succeed if they could prove that few (if any) decisions or acts of e-consumers would intervene in the sequence of events leading to damage.[434] The proof that e-consumers were unable to protect themselves from loss and inevitable damage would help. In the typical TMO third-party liability case, an e-consumer relies on the trustmark and transacts with the e-merchant (as a purchaser of real estate at the lower level of the market relies on the valua-

---

[432] See *supra* Subsection 5.2.2, *Exclusion and limitation of liability clauses*.

[433] See in Section 4 *supra* the proximity factor: defendant's 'assumption of responsibility'.

[434] See Witting, C. (2004), p. 224.

tion and buys the house; or an investor relies on an audit report and buys shares of the audited company). Subsequently, as a result of the undetected (or unreported) inadequate security of the e-merchant's system, or of the unfair e-merchant's privacy policy or business practice, the e-consumer suffers damage which may range from violation of the e-consumer's privacy and data protection rights to pure economic loss (as a purchaser of a house suffers damage for undetected defects of the house; or the investor suffers damage for undetected financial problems of the audited company). In fact, for example: (a) the e-consumer did not receive the good or the service he had paid for; (b) without prior consent from the e-consumer, the e-merchant processed his personal data for purposes other than the fulfilment of the relevant contractual obligations (e.g., used for profiling- and marketing-related purposes, shared with or sold to third parties), with the e-consumer eventually receiving unsolicited marketing e-mails and phone calls from the e-merchant or third parties; (c) the e-consumer's payment details were used directly by the e-merchant to defraud the e-consumer, were shared or sold by the e-merchant to third parties to ultimately defraud the e-consumer, or stolen during the transaction or from the e-merchant's client database by a cunning third party that took advantage of the poor security of the e-merchant's IT infrastructure.[435] Between the TMOs' negligent performance and the subsequent damage to e-consumers is just a 'click'. Moreover, there is evidently no chance for e-consumers to protect themselves from loss and inevitable damage. As a matter of fact, e-consumers cannot intervene in the e-merchant security measures, privacy or business practice.

In conclusion, this analysis suggests that proximity between TMOs and third-party e-consumers may be established under the 'extended Hedley Byrne liability' rule. In fact, it is possible to prove the existence, at some stages prior to the TMOs' failure to take care, of a significant causal link by which TMOs' failure could have resulted in loss to e-consumers, so as to assist the court in identifying TMOs as the subjects with substantial ability to cause loss to e-consumers.

6.2 Validity of TMOs' third-party liability disclaimers

In *Smith* v. *Eric S Bush* and *Harris* v. *Wire Forest District Council*, the House of Lords held that disclaimers were unreasonable and thus contrary

[435] See the typical TMO third-party liability case, Chapter 1, Section 2 *supra*.

to the provisions of the UCTA. As a consequence, given the high degree of proximity between parties, the defendants were liable for their negligent misstatements.[436] Therefore, in order to assess whether the TMOs' liability to e-consumers exists, there is a need to check whether or not the TMOs' disclaimers are reasonable.

TMOs, through contractual clauses or notices,[437] tend to waive their liability for any loss incurred by e-consumers for relying on the trustmarks. Usually, a contract exists only between TMOs and e-merchants. However, common law rules do not always exclude the possibility that the effects of a contract may extend to third parties.[438] Moreover, TMOs try sometimes to place those disclaimers in what they define as Relying Party Agreements or Third Party Agreements, which are meant to impose on e-consumers (which may be defined as 'relying parties' or 'third parties') the TMOs' disclaimers on third-party liability. Relying Party Agreements are posted on TMOs' websites but are not very easy to find. They are practically impossible for e-consumers to read, understand, and enter into; thus, Relying Party Agreements may be compared more with notices than with contracts. Non-contractual notices (e.g., legal notices, security notices, privacy notices) are another means of TMOs to exclude or limit their third-party liability.[439]

In any case, the validity of the TMOs' disclaimers must be assessed. The UCTA regulates a range of situations in which civil liability is excluded or limited.[440] Both contractual clauses and notices are subject to the reasonableness requirement set forth in Section 2 (2) of the UCTA.[441] In deter-

---

[436] See *supra* Subsection 5.2.2, *Exclusion and limitation of liability clauses*.

[437] E.g., Terms and Conditions, Certification Practice Statement, Seal Licence Agreement, Relying Party Agreement.

[438] Courts have indicated at times the need to accommodate the contractual arrangements by which risks of liability have been allocated, so that third parties could take the benefit of a contractual term or to be, in effect, burdened by it. See Powell, J.L. & Stewart, R. (2007), pp. 142-144; Witting, C. (2004), pp. 439 et seq.

[439] See Chapter 3, Subsection 5.5 *supra*.

[440] See also the related provision of the Unfair Terms in Consumer Contract Regulations, Statutory Instrument 1999/2083. The view that has been taken is that these provisions are unlikely to catch any exclusion or limitation clause which has not been rendered invalid by the UCTA. See Powell, J.L. & Stewart, R. (2007), pp. 140-143.

[441] Section 2. 'Negligence Liability. (1) A person cannot by reference to any contract term or to a notice given to persons generally or to particular persons exclude or restrict his

mining whether or not exclusion or limitation clauses or notices are valid, the courts will take an 'in substance' approach, as the House of Lords did in *Smith* v. *Eric S Bush* and *Harris* v. *Wire Forest District Council*. In other words, any attempt to narrow ordinary boundaries of liability must be reasonable. Most of the arguments used by the House of Lords in Smith v. Eric S Bush to reject the defence of the Council for valuers may actually apply to the typical TMO third-party liability case.[442] It is certainly possible to recognise in the Council for valuers' defence of the fairness and reasonableness of the exclusion clause a possible TMOs' defence of their third-party liability disclaimers. In fact, TMOs will also put forward arguments such as:

a) The disclaimers are both clearly written and understandable and are strongly brought to the e-consumers' attention by, e.g., capitalising the letters of the text of the disclaimers;
b) e-consumers cannot rely on the information in the trustmarks;
c) e-consumers should obtain and pay for their own audit of the e-merchant security, privacy or business practice;
d) If TMOs cannot disclaim their liability to third-party e-consumers, they will face a flood of claims, some of which will be unmeritorious but difficult and expensive to resist;
e) TMOs need to be much more cautious in providing trustmarks; the cost of carrying out the audit of e-merchant security, privacy, or business practices will be extremely high, not to mention the cost of negligence insurance for TMOs. All together, the costs will probably be unbearable for TMOs. Consequently, the price of trustmarks will be too high and TMOs will probably exit the market.

The courts, on the other hand, may argue for the inconsistency of TMOs, pursuant to the UCTA, by saying that, although there is no evidence that 90% of e-consumers rely on trustmarks, the TMOs' business model is based

---

liability for death or personal injury resulting from negligence. (2) In the case of other loss or damage, a person cannot so exclude or restrict his liability for negligence except insofar as the term or notice satisfies the requirement of reasonableness. (3) Where a contract term or notice purports to exclude or restrict liability for negligence, a person's agreement to or awareness of it is not of itself to be taken as indicating his voluntary acceptance of any risk.'

[442] See *supra* Subsection 5.2.2, *Exclusion and limitation of liability clauses*.

on e-consumers' reliance on trustmarks. In fact, if e-consumers do not rely on trustmarks, e-merchants will not request those (trustmarks); consequently, there will be no reason for TMOs to exist. Therefore, TMOs know that if the trustmarks' service is inadequately performed, plaintiff e-consumers are very likely to suffer loss.[443]

Furthermore, e-consumers do not usually plan to spend more money than the amount necessary to complete the transaction. The cost of a second audit of the e-merchant security, privacy or business practices will be unbearable in terms of time and cost. All together thus, a second audit is unfeasible for e-consumers who usually visit several websites in a day and purchase goods of modest value. In fact, this will clash with the successful formula of e-commerce, which is: *e-commerce = fast + cheap transactions*. Therefore, it seems unfair and unreasonable to ask e-consumers not to rely on trustmarks and to carry out a second audit of e-merchant security, privacy or business practices at their expense.

Moreover, TMOs are TTPs,[444] which are, by definition, independent and trustworthy professionals. Last but not least, TMOs advertise themselves as independent and trustworthy professionals through massive marketing operations aimed at inducing e-consumers to rely on their trustmarks by creating a fiduciary relationship with them.[445] This may even be seen as the TMOs' *de facto* assumption of responsibility towards e-consumers.[446]

More generally, on the one hand, the e-consumers' bargaining power is much less than the e-merchants', if it is not inexistent. In fact, there is no bargaining activity between the parties. Disclaimers are just unilaterally drafted by TMOs. Additionally, other sources of advice are available only at a time and cost that is unthinkable in an e-commerce transaction. On the other hand, the task undertaken by TMOs is a difficult one.[447] Probably, no insurance company will insure TMOs if they cannot disclaim their liability for negligent misstatement. And, if disclaimers will be considered invalid, TMOs may face a flood of third-party claims.

443 See Subsection 6.1 *supra*.

444 See Chapter 1, Section 1 *supra*.

445 See Chapter 2, Subsection 6.2 *supra*.

446 See *supra* in Section 4 the proximity factor: defendant's 'assumption of responsibility'.

447 See *supra* Chapter 3 and *infra* Chapter 9, Subsection 3.2.3, *Bai-boom trust stage*.

In conclusion, there is no clear-cut answer to the question on the validity of TMO third-party liability disclaimers. A number of arguments for and against the unfairness of the disclaimers have been unfolded. In the end, the courts will most likely decide on the basis of policy arguments.[448]

6.3 Policy arguments

There are two main policy considerations highlighted in the 'extended Hedley Byrne liability' line of cases. The first, and more general one, is that the courts saw the imposition of a third-party duty of care on the defendant as fair, just, and reasonable on the basis that the defendants were performing socially important functions which eventually exposed the plaintiff to the harm in question. The second one, which was specific to *Smith* v. *Eric S Bush* and *Harris* v. *Wyre Forest District Council*, is that it was not reasonable to expect a purchaser of a house at the lower level of the market to obtain an independent valuation.[449]

However, as already argued, policy arguments are really dependent on the circumstances and thus have to be dealt with case by case. As there has been no case of a TMO third-party liability, after having dealt with the two specific policy arguments of the 'extended Hedley Byrne liability' line of cases, a number of more general policy arguments will be tested in the typical TMO third-party liability case in order to anticipate related court arguments.

6.3.1 *Do TMOs discharge a socially important function?*

In the 'extended Hedley Byrne liability' line of cases, duties of care are imposed on the basis that the defendant was performing a socially important function. Whether TMOs perform a socially important function is up to the courts to decide. It is arguable that e-consumer trust is crucial to reaping the full benefit of e-commerce.[450] E-consumers are not capable of scrutinising e-merchants' policies. Ideally, through trustmarks, e-consumers may

[448] For more information on e-consumer protection in information technology contracts in England, see Girot, C. (2000), pp. 21-24.

[449] See *supra* Subsection 5.2.3 *Policy arguments*.

[450] See Chapter 1, Section 1 *supra*.

receive a sort of guarantee from TMOs (i.e., independent third parties) of the quality of the e-merchant security, privacy or business practice. Moreover, trustmarks are easy to recognise and may improve e-consumers' perception of potential online 'business partners', provide 'always available', 'independent', and trustworthy information on e-merchants and thus enhance e-consumers' trust in online transactions. Given this, it can be said that the TMOs perform a socially important function. Consequently, if TMOs negligently fail to provide their services, it will be fair, just, and reasonable to impose on them a duty of care.

6.3.2 *Is it reasonable to require that e-consumers obtain an independent evaluation?*

The argument used by the court in Smith v. Eric S Bush and in Harris v. Wyre Forest District Council cannot be directly applied to the typical TMO case. In fact, as a purchaser of a house pays for the costs of the valuation conducted on behalf of the mortgagee, an e-consumer does not pay directly for the trustmark. Nevertheless, it is also unreasonable to require e-consumers to obtain an independent evaluation of e-merchant security, privacy, or business practice. In fact, it is simply ridiculous to expect that e-consumers who usually visit several websites in a day and purchase goods of modest value will obtain an independent assessment of the security, privacy, or business practices of all e-merchant websites in order to decide whether or not to transact with them. Therefore, the unreasonableness of requiring e-consumers to obtain an independent evaluation seems undisputed.

6.3.3 *Floodgates arguments*

Floodgates and fear of indeterminacy are the most recurrent policy arguments related to third-party liability for negligent misstatement. They are definitely applicable to the TMO third-party liability for negligently issued trustmarks. In fact, the worldwide accessibility of trustmarks in e-merchant websites creates the possibility that an indefinite number of e-consumers will potentially rely on the trustmark. Allowing a cause of action against TMOs for aggrieved e-consumers who relied on trustmarks would be like opening the floodgates. Meaning, defendant TMOs would be flooded by claims, possibly resulting in their financial ruin. Courts could adopt two approaches towards floodgates arguments. They could 'close' the floodgates by maintaining that TMOs should be in a position where they are able to weigh the costs of taking precautions against the possible size of claims that could be made against them and that TMOs should be able to predict the number of persons that their negligence might affect. Evidently, TMOs are neither able to estimate the cost of claims they might face nor predict the number of persons that their negligence might effect. The courts could conclude that it would not be fair, just, and reasonable to impose a duty of care towards e-consumers and deny them a cause of action.

However, the House of Lords had already approached the floodgates concern with suspicion,[451] expressing the view that the '[d]enial of the existence of a cause of action is seldom, if ever, the appropriate response to fears of its abuse.'[452] In this respect, in *Smith* v. *Eric S Bush* and in *Harris* v. *Wyre Forest District Council*, a cause of action for purchasers of real estate at the lower level of the market against valuers was allowed. Nevertheless, valuers' exposure to claims was far less than the TMOs'. In fact, the valuation, as a paper-based document, has a limited and rather slow diffusion. Therefore, it can be seen and relied upon by potential purchasers of the house in that region of England. Trustmarks, as electronic documents available online, can be seen and relied upon worldwide; hence, claims against TMOs may come from all around the world. Moreover, as it will be explained in the next section, the English legislator has already codified the existence of a cause of action in case of third-party liability for the negligent issuance of digital certificates against CSPs.[453] In fact, CSPs are TTPs which are exposed to the same risk of claims as TMOs. The digital certificate is, in fact, attached to all signatory signatures and is available worldwide on the Certification Revocation List (CRL).

---

[451] See, e.g., *McLoughlin* v. *O'Brian* [1983] 1AC 425, HL.
[452] See, e.g., *Phelps* v. *Hillingdon London BC* [2001] 2 AC 667, HL.
[453] See Section 4 of the Electronic Signatures Regulation 2002.

7. Parallel TMOs-CSPs: applicability by analogy of CSP third-party liability rules to TMOs

Next to the option of comparing TMOs with surveyors, auditors, and accountants, TMOs can be compared with CSPs.[454] The third-party liability of CSPs is set out in Article 6[455] of the Electronic Signatures Directive.[456] England implemented Article 6 of the Electronic Signatures Directive in Section 4 of the Statutory Instrument 2002 No. 318 (henceforth: The Electronic Signatures Regulation 2002), as follows:

> 'Liability of certification-service-providers 4.
> (1) Where – (a) a certification-service-provider either – (i) issues a certificate as a qualified certificate to the public, or (ii) guarantees a qualified certificate to the public, (b) a person reasonably relies on that certificate for any of the following matters – (i) the accuracy of any of the information contained in the qualified certificate at the time of issue, (ii) the inclusion in the qualified certificate of all the details referred to in Schedule 1, (iii) the holding by the signatory identified in the qualified certificate at the time of its issue of the signature-creation data corresponding to the signature-verification data given or identified in the certificate, or (iv) the ability of the signature-creation data and the signature-verification data to be used in a complementary manner in cases where the certification-service-provider generates them both, (c) that person suffers loss as a result of such reliance, and (d) the certification-service-provider would be liable in damages in respect of any extent of the loss – (i) had a duty of care existed between him and the person referred to in subparagraph (b) above, and (ii) had the certification-service-provider been negligent, then that certification-service-provider shall be so liable to the same extent notwithstanding that there is no proof that the certification-service-provider was negligent unless the certification-service-provider proves that he was not negligent.
> (2) For the purposes of the certification-service-provider's liability under paragraph (1) above there shall be a duty of care between that certification-service-provider and the person referred to in paragraph (1) (b) above.
> (3) Where – (a) a certification-service-provider issues a certificate as a qualified certificate to the public, (b) a person reasonably relies on that certificate,

[454] See Chapter 1, Subsection 3.1 *supra*.

[455] Article 6. Liability, see *supra* n. 32.

[456] For an in-depth analysis of CSP liability as set forth in Article 6 of the Electronic Signatures Directive, see Balboni, P. (2004), pp. 211-242.

(c) that person suffers loss as a result of any failure by the certification-service-provider to register revocation of the certificate, and (d) the certification-service-provider would be liable in damages in respect of any extent of the loss – (i) had a duty of care existed between him and the person referred to in sub-paragraph (b) above, and (ii) had the certification-service-provider been negligent, then that certification-service-provider shall be so liable to the same extent notwithstanding that there is no proof that the certification-service-provider was negligent unless the certification-service-provider proves that he was not negligent.
(4) For the purposes of the certification-service-provider's liability under paragraph (3) above there shall be a duty of care between that certification-service-provider and the person referred to in paragraph (3) (b) above.'

For our analysis, it is important to stress that a CSP is liable to a person who reasonably relies on the CSP's certificate – and suffers loss as a result of such reliance – for the accuracy of the information contained in the certificate at the time of issue, notwithstanding the absence of proof of the CSP's negligence, unless the CSP proves it was not negligent (see Section 4 (1)). The English legislator sets forth that CSPs have the duty to provide certificates containing information which is accurate at the time of issue. Moreover, the English legislator operates a presumption of the CSPs' liability towards third parties who suffer loss as a result of their reasonable reliance on the CSPs' certificates. However, CSPs could prove that they were not negligent.

As Section 4 of the Electronic Signatures Regulation is the only statutory provision in England on the liability of a TTP, it is possible that the courts will take it into consideration once they have to decide on the liability of other TTPs (e.g., TMOs). If this is the case, TMOs will be liable to e-consumers who reasonably rely on TMOs' trustmarks – and suffer loss as a result of such reliance – for the accuracy of the information contained in the trustmark at the time of issue, notwithstanding the absence of proof of the TMOs' negligence, unless the TMOs prove that they were not negligent. The TMOs' standard of care will thus consist of providing accurate information on e-merchants in their trustmarks. Aggrieved e-consumers who reasonably rely on TMOs' trustmarks may ask compensation from the TMOs for the latter's negligent performance, without having to prove such negligence. The TMOs will be held liable unless they prove they were not negligent.

It is then crucial at this point to establish reasonability of the e-consumers' reliance. This is a proximity question.[457] In this respect, it could be argued that the e-consumers' reliance is reasonable, as in a typical TMO third-party liability case, in which the proximity factors are all satisfied.[458]

Moreover, it should be noted that in the implementation of Article 6 of the Electronic Signatures Directive, the English legislator left out the chance for CSPs to limit their liability towards third parties. In fact, the wording of Article 6 (3) and (4)[459] cannot be traced in Section 4 of the Electronic Signatures Regulation 2002. This omission may be used as an extra argument to consider the TMO third-party liability disclaimer unreasonable[460] and thus to strengthen the proximity relation between TMOs and third-party e-consumers.

8. Conclusions

Given the absence of case law and statutory instruments on the third-party liability of TMOs, the central question that has been addressed in this chapter is whether it is reasonable to impose on TMOs a duty of care towards e-consumers; or, looking at the issue from another angle, whether e-consumers who detrimentally relied on trustmarks and suffered loss as a consequence have a cause of action against TMOs that negligently performed their services.

It has been pointed out that among the three elements of the duty of care, matters of proximity and policy arguments hold the key to duty issues in which misstatement cases are to be analysed whereas foreseeability is of minor importance in the present analysis.[461] A proximity relation between

---

[457] See Subsection 4.2 *supra*.

[458] See Subsection 5.2 *supra*.

[459] Article 6: '(...) 3. Member States shall ensure that a certification service provider may indicate in a qualified certificate limitations on the use of that certificate, provided that the limitations are recognisable to third parties. The certification service provider shall not be liable for damage arising from use of a qualified certificate which exceeds the limitations placed on it. (4). Member States shall ensure that a certification service provider may indicate in the qualified certificate a limit on the value of transactions for which the certificate can be used, provided that the limit is recognisable to third parties. The certification service provider shall not be liable for damage resulting from this maximum limit being exceeded.'

[460] See Subsection 5.2 *supra*.

[461] See Section 5 *supra*.

TMOs and e-consumers seems to be established under the 'extended Hedley Byrne liability' rules.[462] Policy arguments will eventually play a crucial role in the courts' decisions on whether or not it is reasonable to impose on TMOs a duty of care towards e-consumers and thus allow a cause of action for aggrieved e-consumers against TMOs that negligently provided wrongful information on e-merchants. More precisely, the answer to the central question addressed in this chapter will most likely depend on the approach that courts will take to floodgates arguments.[463] Courts usually tend to 'close' the floodgates, trying to avoid potential liability to an unlimited number of people for an indeterminate amount. However, the rule set out in Section 4 of the Electronic Signatures Regulation 2002 imposes on CSPs a duty of care towards users who rely on the information contained in the electronic certificate. It has been observed that CSPs are TTPs comparable with TMOs.[464] This rule may either be applied by analogy to TMOs or considered by the courts as a tendency of the English legislator to impose a third-party duty of care on TMO-like professionals. Courts may thus allow a cause of action for aggrieved e-consumers against TMOs that negligently provide their services.

[462] See Subsection 6.1 *supra*.
[463] See *supra* Subsection 6.3, *Floodgates arguments*.
[464] See Chapter 1, Subsection 3.1 *supra*; and Section 7 of the present chapter *supra*.

Chapter 6
GERMANY

1. Introduction

As in England, neither case law nor literature has specifically addressed the issue of TMO third-party liability in Germany. Neither is there a trace of specific statutory provisions on the matter. This chapter thus aims to determine whether e-consumers who relied on inaccurate trustmarks and suffered loss as a consequence could ask TMOs for compensation either in general tort or contract law.

Section 2 explains the overlap between tort and contract law in matters related to third-party professional liability for negligent misstatements and deals with the relevant tort and contract provisions. In Section 3, the applicability of the relevant tort and contract provisions to a typical TMO third-party liability case is analysed. A selection of Federal Supreme Court decisions potentially applicable to TMOs is offered in Section 4. The applicability and the possible impact of the selected case law on TMO third-party liability is explained in Section 5. Section 6 discusses the applicability by analogy of CSP third-party liability rules to TMOs. Section 7 contains the conclusions.

2. Third-party professional liability for negligent misstatement: the thin line between tort and contract law

At first glance, the German legal system does not seem to offer much ground to grant compensation for pure economic loss caused by negligent statements.[465] The two general tort liability clauses: Sections 823[466] and 826[467] of the German Civil Code (*Bürgerliches Gesetzbuch*, henceforth: BGB) grant protection against (a) injury to life, health, property, or 'other rights' – varying from right to privacy and data protection to freedom of speech – (Section 823 (1)); (b) violation of 'protective norms' – provision protecting individual interests and assets – (Section 823 (2));[468] and (c) economic loss but only when it is caused by an intentional conduct *contra bonos mores* (*Sittenverstoß*). Accordingly, pure economic loss caused by inaccurate information/statements provided by professionals on whom third parties had relied does not enjoy protection under Sections 823 and 826.[469]

The documented gap stimulated a debate, across the 1970s and 1980s, on the possible extension of tort liability in order to protect the aforementioned instances of economic loss.[470] Three main theories were developed. The first was presented by Christian von Bar, who proposed to create the brand-new Section 828 BGB, which would have set forth tort liability for damage caused in a business relationship (*Schädigungen im geschäftlichen Verkehr*) by negligent false information or advice provided by a professional who, for the task he undertakes and the status he enjoys, is especially

[465] See Van Dam, C. (2006), pp. 75-76. For an overview of the German approach to third-party professional liability for negligent misstatements, see Zweigert K. & Kötz, H. (1998), pp. 598 et seq.; Markesinis, B.S. & Uberath, H. (2002), pp. 57 et seq. For more details on pure economic loss, see Van Dam, C. (2006), pp. 172-173; Magnus, U. (2001), p. 95.

[466] Section 823 – 'Duty to compensate for damage: (1) A person who, wilfully or negligently, unlawfully injures the life, body, health, freedom, property or other right of another is bound to compensate him for any damage arising therefrom. (2) The same obligation is placed upon a person who infringes a statute intended for the protection of others. If, according to the provisions of the statute, an infringement of this is possible even without fault, the duty to make compensation arises only in the event of fault.'

[467] Section 826 – 'Wilful damage contrary to public policy: A person who wilfully causes damage to another in a manner contrary to public policy is bound to compensate the other for the damage.'

[468] E.g., in case of fraud, a claim for pure economic loss may be based on Section 823 (2) BGB, in connection with Section 263 of the German Criminal Code (Strafgesetzbuch – StGB). For an extensive discussion on the matter, see Van Gerven, W. et al. (2000), pp. 227-228.

[469] See Markesinis, B.S. & Uberath, H. (2002), pp. 57-64; Van Dam, C. (2006), pp. 172-173.

[470] See Schlechtriem, P. (1998), pp. 13-15.

trusted by third parties.[471] The second theory was put forward by Konrad Huber, who maintained that professionals' duty of care to persons who rely on the information they supply should be considered a 'protective norm' under Section 823 (1) BGB. Auditors, consultants, lawyers and other professionals who provide informed statements would accordingly be held liable for pure economic loss caused to third parties who had relied on inaccurate information which they had possibly supplied.[472] The third theory was proposed by Claus-Wilhelm Canaris, who observed that third-party reliance on presumed accurate information was the very reason for economic loss, the common feature of negligent misrepresentation cases. Therefore, he elaborated a theory of professionals' responsibility for 'reliance loss'.[473] None of the three theories ever found their way into case law.[474]

Nevertheless, the need to protect the economic interests of third parties was unquestionable.[475] Therefore, courts found a way to protect aggrieved third parties through contract law, assuming the existence of a contract between the professional who provided information and the third party that relied on it.[476] These contractual or quasi-contractual means used by the courts are as follows: the contract with protective effects towards third parties (*Vertrag mit Schutzwirkung zugunsten Dritter*, henceforth: *VmSzD*); the implied contract for the provision of correct information (*stillschweigend geschlossener Auskunftsvertrag*, henceforth: *sgA*); and the contractual principle of *culpa in contrahendo*, the principle that parties to a relationship akin to a contract have taken each other's interests into consideration. *VmSzD* and *sgA* are the ones used mostly by courts in decisions on third-party professional liability.[477]

The phenomenon of courts creating contractual construction to overcome the absence of tortuous remedies for economic loss was significantly

---

471 See Von Bar, C. (1981), p. 1761.

472 See Konrad, H. (1978), p. 359 et seq.

473 See Canaris, C.W. (1971).

474 See Schlechtriem, P. (1998), pp. 14-15.

475 See Hirte, H. (1996) *Berufshaftung* (Munich: C.H. Beck).

476 See Van Dam, C. (2006), pp. 75-76; Zweigert K. & Kötz, H. (1996), p. 617; BGH 23.1.1985 in *JZ* 1985, pp. 951 et seq., noted by Honsell, 'Zur Auskunftsvertrag mit Schutzwirkung für Dritte'; Canaris, C. (1983), pp. 567 et seq.; Kreuzer, K.F. (1976), p. 778; Lorenz, W. (1973), pp. 575 et seq.

477 See Markesinis, B.S. & Uberath, H. (2002), pp. 57-64 and pp. 265-295.

described by John Fleming as: 'the German willingness, indeed eagerness, to extend tort protection…[despite] the Civil Code's categorical exclusion of tort damages for pure economic loss and the great weight reputedly given by German law to theoretical orthodoxy over pragmatism.'[478]

After this brief excursus, it is already clear that third-party professional liability lies on the borderline between contract and tort. Furthermore, third parties' reasonable reliance on the negligent misstatement provided by professionals seems to play a crucial role in the compensation of pure economic loss. The following questions seem relevant to the present analysis:

Can protection in tort be completely ruled out?

How do the quasi-contractual constructions work?

Under which conditions may third-party reliance assumed to be reasonable?

In order to provide informed answers, the relevant tortious and contractual provisions need to be dealt with in greater detail.

2.1 Analysis of the relevant tort provisions

Section 823 (1) BGB
'A person who, wilfully or negligently, unlawfully injures the life, body, health, freedom, property or other right of another is bound to compensate him for any damage arising therefrom.'

The present provision sets forth the criterion for the compensation of damages caused by a negligent, unlawful conduct. Accordingly, only a plaintiff who suffers damages related to the breach of one of the legal interests specified – life, body, health, freedom, ownership or any other right of another – can recover them under Section 823 (1) BGB. It is worth stressing for the sake of the present analysis that the concept of 'other right[s]' embraces in principle all the interests protected according to the German legal system *erga omnes*, which has also included, since recently, the right to privacy and data protection.[479] As to economic loss, the rule is that a plaintiff who

[478] Fleming, J. (1989), p. 508; Van Dam, C. (2006), pp. 172-173.
[479] Bassenge P. et al. (2006), p. 1248; Van Gerven, W. et al. (2000), p. 142.

suffered economic loss as a consequence of a defendant's negligent conduct, breaching one of the plaintiff's legal interests specified in the provision, or 'other right[s]', may recover it. Otherwise, there would be no ground for a claim for 'pure' economic loss unrelated to any of such breaches of protected interests or rights.[480]

> *Section 823 (2) BGB*
> 'The same obligation is placed upon a person who infringes a statute intended for the protection of others. If, according to the provisions of the statute, an infringement of this is possible even without fault, the duty to make compensation arises only in the event of fault.'

The second paragraph of Section 823 BGB, which naturally has to be read in connection with the first paragraph, extends the protection of damages suffered by negligent conduct. Liability is expanded to the defendant's actions or omissions which infringe specific rules set forth in statutes protecting other persons' rights or interests. In fact, in the German legal system, compensation for 'pure' economic loss may be sought only under Section 823 (2) BGB or, as it will be explained further below, pursuant to Section 826 BGB. More precisely, Section 823 (2) is the only provision of the BGB which allows the plaintiff's recovery of pure economic loss caused by the defendant's negligent conduct. Accordingly, there must be a special legitimating element in a specific statute for Section 823 (2) to apply.[481] In other words, a claim under this provision will arise only if the damage resulted from the very violation of the rights and interests that a statute aims to protect. For example, if the statute sets forth rules for the protection of personal injuries or damage to properties only, pure economic loss may not be claimed pursuant to Section 823 (2) BGB.[482]

> *Section 826 BGB*
> 'A person who wilfully causes damage to another in a manner contrary to public policy is bound to compensate the other for the damage.'

---

[480] See Markesinis, B.S. & Uberath, H. (2002), pp. 52 et seq.; Van Gerven, W. et al. (2000), pp. 227-228.

[481] See Larenz, K. & Canaris, C.W. (1994), pp. 436-437.

[482] See Van Gerven, W. et al. (2000), p. 228; Larenz, K. & Canaris, C.W. (1994), pp. 435 and 439.

As mentioned right before, compensation for pure economic loss may be sought pursuant to Section 826 BGB. There are two requirements for a claim for pure economic loss under Section 826 BGB to succeed: the defendant must have acted *contra bonos mores* (*Sittenwidrig*) and with intent (*Vorsatz*). In contrast to Section 823, the defendant's conduct does not have to violate selected plaintiff's rights and interests or specific statutes. In other words, this general clause grants protection to violations of all rights and interests, pure economic loss included, provided that such violations are *contra bonos mores* and carried out with intent.

Some words need to be spent on the mentioned requirements.

A conduct is *contra bonos mores* when it offends the 'fundamental concepts of morally acceptable conduct towards persons with whom one is in a legal relationship.'[483] Furthermore, *boni mores* entail a set of legal-ethical principles which changes with the developments of the society they refer to. For example, standards of proper economic conduct in commercial practices can fit into the concept of *boni mores*. Recently, it has been widely accepted that 'acting *contra bonos mores* implies conduct contrary to the existing economic and legal order or the *ordre public*.'[484] In this respect, providing someone with incorrect information is a type of conduct which falls within the scope of Section 826 BGB.[485]

Turning now to the concept of intent, it has a broader meaning than its literal one. It refers not only to the 'will to harm' but also to deception (*Täuschung*) and recklessness (*Leichtfertigkeit*).[486] Accordingly, the requisite of intent is not only fulfilled when 'the defendant actually intended to cause the harm; it is enough if he was conscious of the possibility that harm might occur and acquiesced in its doing so.'[487] In other words, it is sufficient that the defendant is aware of the possible harmful consequences of his conduct but nevertheless accepts them as inevitable, even without desiring them. (i.e., *dolus eventualis*).[488] A professional who, for example, pro-

---

[483] *Grundanschauungen loyalen Umgangs unter Rechtsgenossen*: BGH, 2 June 1981, *NJW* 1981, p. 2184 at p. 2185.

[484] Van Dam, C. (2006), pp. 401-402; see also Markesinis, B.S. & Uberath, H. (2002), pp. 889-890; Larenz, K. & Canaris, C.W. (1994), p. 448.

[485] See Van Dam, C. (2006), p. 402.

[486] Larenz, K. & Canaris, C.W. (1994), pp. 452-453.

[487] Zweigert K. & Kötz, H. (1998), p. 603.

[488] See Van Dam, C. (2006), pp. 71 and 172; Markesinis, B.S. & Uberath, H. (2002), pp. 292 and 889.

vides inaccurate information on the creditworthiness of another party, being aware that third parties who will possibly rely on such information may as a consequence suffer loss, could be held liable pursuant to Section 826 BGB.[489] The requirement of intent has thus been expanded to cover thoughtless acts in cases where the risk of causing loss to third parties has been accepted.[490]

Evidently, Section 826 BGB plays a crucial role in third-party professional liability; it is indeed used quite often in cases of negligent misstatement.[491]

2.2 Analysis of the relevant contractual or quasi-contractual provisions

Given the letter of Section 676 BGB: '[a] person who gives advice or a recommendation to another is not bound to compensate for any damage arising from following the advice or recommendation, without prejudice to his responsibility arising from contract or tort' and the manifest difficulty in recovering pure economic loss in tort; courts have created ways to claim pure economic loss in contract.[492] More precisely, in cases of third-party liability for negligent misstatement, courts operate a sort of 'fiction' to set up a contract. The basic idea is that courts would consider requests for information or advice as offers – to execute a contract for the provision of information or advices – and the positive answer of the professionals (e.g., auditors, accountants, surveyors, banks) as an acceptance.[493] This way, a

---

489 However, it must be noted that gross negligence (i.e., an objectively gross deviation from reasonably careful conduct) does not in itself qualify as wilfulness for the purposes of Section 826, although in practice 'lesser' forms of wilfulness such as recklessness may very well cover gross negligence cases. See Larenz, K. & Canaris, C.W. (1994), pp. 454-455.

490 See Von Bar, C. (1994), pp. 103 et seq.

491 See Markesinis, B.S. & Uberath, H. (2002), pp. 890-891.

492 See Van Dam, C. (2006), pp. 75-76; Zweigert K. & Kötz, H. (1996), p. 617; BGH 23.1.1985 in *JZ* 1985, pp. 951 et seq., noted by Honsell, 'Zur Auskunftsvertrag mit Schutzwirkung für Dritte'; Canaris, C. (1983), pp. 567 et seq.; Kreuzer, K.F. (1976), p. 778; Lorenz, W. (1973), pp. 575 et seq.

493 E.g., when a bank gave information about the creditworthiness of one of its clients to another person (BGH *NJW* 1972, p. 1200.), when a finance newsletter recommended certain investments (*BGHZ* 70, 356; *BGHZ* 74, 103, 107), or when an architect gave information to the mortgage bank about the progress of the construction of a building (OLG Hamm, *ZfBR*

quasi-contract (so defined by the main doctrine[494]) is concluded by and between professionals and the party who has asked for information or advice. From the basic idea, courts have developed quite far-reaching quasi-contractual frameworks, such as *VmSzD*, *sgA*, and *culpa in contrahendo* – the principle that parties to a relationship akin to a contract have, to some extent, taken each other's interests into consideration, in order to compensate pure economic loss suffered from persons who had relied on the information or advice, also in cases where they did not ask for them. How *VmSzD*, *sgA* and *culpa in contrahendo* exactly work is explained hereunder.

2.2.1 *Contract with protective effects towards third-parties* (*Vertrag mit Schutzwirkung zugunsten Dritter – VmSzD*)

Since 1965, the *VmSzD* has been used to compensate third-party economic loss caused by professional negligence.[495] This has been defined as a judicially created variant of the contract for the benefit of third parties (*Vertrag zugunsten Dritter – VzD*).[496] The *VzD* is regulated by Sections 328 ss. BGB.[497] Accordingly, a subject may wilfully oblige himself, by means of a contract with another subject, to undertake a performance to a third party. The consequence of this action is that 'the third party directly acquires the right to demand the performance' (Section 328 (1)).[498] A major difference between the *VzD* and the *VmSzD* rests in the fact that in the latter, the intent of the professional to take on contractual obligations towards third parties is assumed by the courts. In fact, if the professional who provides information will be asked, he will most likely deny his will to include third parties as beneficiaries of the contractual performance.[499] Moreover, the third party

1987, 42), and such information later turned out to be inaccurate, the losses suffered by those relying on it were made recoverable by assuming the formation of a contract. See Lorenz, W. (1973), pp. 575 et seq.; see also BGH 23.1.1985 in *JZ* 1985, pp. 951 et seq., noted by Honsell, 'Zur Auskunftsvertrag mit Schutzwirkung für Dritte'.

[494] See, e.g., Markesinis, B.S. & Uberath, H. (2002), p. 58; Schlechtriem, P. (1998), p. 14.

[495] BGH *NJW* 1965, 1955. For a more general view of the important role played by *VmSzD* in third-party professional liability, see Hirte, H. (1996), p. 388.

[496] See Markesinis, B.S. & Uberath, H. (2002), pp. 59-64 and p. 292.

[497] On such provisions, see extensively Bassenge P. et al. (2006), pp. 530 et seq.

[498] For an analysis of the *VzD*, see Zweigert K. & Kötz, H. (1998), pp. 459-460.

[499] See Markesinis, B.S. & Uberath, H. (2002), p. 294. Such interpretation may also be supported by looking at the disclaimers on third-party liability that professionals usually set forth in the contracts on the provision of their services.

does not need to be either specifically identified in advance or somehow in a close relationship with the contractual creditor. To check whether a *VmSzD* can be found, courts generally evaluate if in fact it is possible to infer from the interests involved that the parties have agreed (also implicitly) to establish the duty of the professional providing information towards third parties.[500] Moreover, third parties' (reasonable) reliance on the accuracy of professionals' statements, coupled with the special skills of the latter, may play on a case-by-case basis a significant role. A relevant example is given by 'case 3',[501] in which although the conflicting interests involved did not suggest that the parties had agreed to stipulate for a contractual duty towards third parties, a *VmSzD* was nevertheless found on the basis of reasonable third party reliance.[502]

If, on the one hand, courts are not very consistent on the requisite to grant compensation to third parties under *VmSzD*, on the other hand, the trend to move for an expansion of the 'contractual umbrella' is clear.[503] Such courts' approach poses, however, the issue of how to define and thus limit the group of third parties which may recover pure economic loss under *VmSzD*. The analysis of the relevant case law will help find an answer to this question.[504]

2.2.2 *Implied contract on the provision of correct information (Stillschweigend geschlossener Auskunftsvertrag – sgA)*

An alternative to the *VmSzD*, the *sgA* is used by German courts as a quasi-contractual legal ground to compensate a third party for pure economic loss suffered from relying on inaccurate information provided by professionals. To justify the imposition of third-party liability on a professional under an *sgA*, courts consider that by disseminating information, the professional is aware that it will be relied on by third parties and hence implicitly takes on the related liability. As a general rule, a direct interaction between the pro-

---

500 See BGH *NJW* 1984, pp. 355-356.

501 See Subsection 4.3 *infra*.

502 For the difference between the *VzD* and the *VmSzD*, see extensively Larenz, K. (1956), p. 16. For the condition required by courts for a *VmSzD* to exist, see Beyer, O. (1996), p. 473; Sonnenschein, J. (1989), p. 225.

503 See Markesinis, B.S. & Uberath, H. (2002), pp. 64-65.

504 See Section 4 *infra*.

fessional and the third party to whom he provided the information would help courts assume the existence of an *sgA*. In the absence of such a contact between the parties, courts tend to choose the *VmSzD* over the *sgA* as grounds for third-party professional liability.[505] Moreover, there are a number of elements which play a significant role in the courts' decision on the existence of an *sgA*. In fact, if (a) the professional knew that the information he had provided was significant to a third party's further decisions; (b) the professional prepared and issued the information using his professional expertise; (c) the professional had economic interest in providing such information;[506] (d) the professional issued a warranty on the quality of information provided; (e) the third party used the information within the scope for which it had been issued by the professional;[507] an *sgA* would most likely be found.[508]

2.2.3 *Culpa in contrahendo*

The principle of *culpa in contrahendo*, whose doctrine was developed by the famous German jurist Rudolf von Jhering,[509] is a means to broaden contractual liability. In fact, through this concept, contractual remedies are extended to the pre-contractual phase, i.e., the negotiation phase. For example, when a professional warrants that the contract will most likely have a positive outcome, knowing that the other contracting party will take ac-

---

[505] See Markesinis, B.S. & Uberath, H. (2002), pp. 265-300; Von Bar (1994), p. 111.

[506] Such elements are especially useful to prove the requisite of intent to assume responsibilities towards third parties on the side of the professional in an *sgA*. See the court's wording in the following decision: *die Gesamtumstände unter Berücksichtigung der Verkehrsauffassung un des Verkehrsbedürfnisses den Rückschluß zulassen daß beide Teile nach dem objektiven Inhalt ihrer Erklärungen die Auskunft zum Gegenstand vertraglicher Rechte un Pflichten gemacht haben*'. BGH *NJW* 1992, p. 2080 at p. 2082. See also, on the point, BGH *WM* 1985, p. 1531 at p. 1532; BGH *NJW* 1992, p. 2080 at p. 2082; BGH *NJW* 1998, p. 1244.

[507] See, e.g., BGH *WM* 1986, p. 711.

[508] Quite few similarities may be found with the requirements for establishing 'proximity' under English law. See Chapter 5, Section 4 and Subsection 6.1 *supra*.

[509] See Jhering, R. (1861) Culpa in contrahendo oder Schadenersatz bei nichtigen oder nicht zur Perfektion gelangten Verträgen, in Jhering (ed.) Jahrbücher für die Dogmatik des heutigen römischen und deutschen Rechts 4, pp. 1-112. See also Medicus, D. (1986), *Entdeckungsgeschichte der culpa in contrahendo*, in Festschrift für Max Kaser zum 80. Geburtstag, p. 169.

tions upon such sort of assurance, but eventually the professional's performance falls short in some way of what has been promised, there may be a claim based on *culpa in contrahendo* against the professional.[510] In the perspective of the present analysis, one might think of a professional who, during the negotiations, personally induced the third party's confidence in the trustworthiness of the information provided. However, *culpa in contrahendo* almost does not have any impact on third-party professional liability for negligent misstatements when compared with the role played by *sgA* and *VmSzD*.[511]

3. IMPACT OF TORT AND CONTRACT PROVISIONS ON THIRD-PARTY TMO LIABILITY

Losses caused by inaccurate trustmarks issued by TMOs may not fall under the protection of Section 823 (1) BGB. It is very unlikely that inaccurate trustmarks will cause injury to e-consumers' legal interests, such as life, body, health, freedom, or ownership. However, in the instance that e-consumers suffer damage for breach of their privacy rights, this may fall under the definition of 'other rights' and thus make Section 823 (1) BGB applicable in this specific case.[512] As to Section 823 (2) BGB, it could only apply if the losses suffered by e-consumers who had relied on inaccurate trustmarks are specifically protected by the law. Section 826 BGB might not apply to TMO cases unless the TMO issued the inaccurate trustmark with the intention to cause harm to e-consumers. However, as already pointed out, the 'intent' required by Section 826 BGB has recently been interpreted to also include thoughtless acts and recklessness (*Leichtfertigkeit*).[513] Therefore, it seems enough for the 'intent' element to be fulfilled to be aware of the consequences of one's conduct and to accept them as inevitable even without desiring them (*dolus eventualis*).[514] According to this interpreta-

[510] See BGH *NJW* 1972, p. 1189; BGH *WM* 1967, p. 798; BGH *NJW* 1965, p. 812.

[511] See Canaris (1999), p. 220.

[512] See Subsection 2.1 *supra*; Bassenge P. et al. (2006), p. 1248. Van Gerven, W. et al. (2000), p. 142.

[513] See Subsection 2.1 *supra*; Larenz, K. & Canaris, C.W. (1994), pp. 452-453; Zweigert K. & Kötz, H. (1998), p. 603.

[514] See Van Dam, C. (2006), pp. 71 and 172; Markesinis, B.S. & Uberath, H. (2002), pp. 292 and 889.

tion, if TMOs' negligence is deemed a result of recklessness, e-consumers may have a cause of action against TMOs under Section 826 BGB. However, '*dolus eventualis*' is a necessary but not sufficient condition for Section 826 BGB to apply. In addition to it, TMOs must have been conscious of the risk of damaging third-party e-consumers and have accepted it. Moreover, the TMOs' conduct must be considered *contra bonos mores*, such as a behaviour that significantly offends the fundamental concept of morally acceptable conduct towards persons with whom one is in a legal relationship (e.g., acting against standards for proper economic conduct in commercial practices).[515]

E-consumers, however, may not only have an action against TMOs in tort; they may also seek redress in contract or, to be more precise, in 'quasi-contract'.[516] Quasi-contractual means, such as *VmSzD* and *sgA*, and, to a lesser extent, the contractual principle of *culpa in contrahendo,* may offer a third party who has suffered loss from relying on inaccurate trustmarks some legal ground to recover such loss.[517]

The *VmSzD* seems in principle applicable to the typical TMO case. In fact, TMOs' intent to oblige themselves towards third-party e-consumers could be assumed; e-consumers do not need to be specifically identified in advance or to be in a close relationship with the certified e-merchants; and TMOs are professionals specifically skilled in providing information on e-merchant security, privacy, and business practice. However, it remains to be assessed, case by case, whether the reliance of e-consumers on the trustmarks is reasonable and if the interests involved are such for the courts to infer that TMOs and e-merchants agreed (even implicitly) to establish a duty of TMOs towards e-consumers.[518]

As far as the *sgA* is concerned, there is no direct contact between TMOs and third-party e-consumers. Usually, in this situation, the courts tend to

---

515 Van Dam, C. (2006), pp. 401-402; see also Markesinis, B.S. & Uberath, H. (2002), pp. 889-890; Larenz, K. & Canaris, C.W. (1994), p. 448.

516 See Subsection 2.2 *supra*; see also Van Dam, C. (2006), pp. 75-76; Zweigert K. & Kötz, H. (1996), p. 617; BGH 23.1.1985 in *JZ* 1985, pp. 951 et seq., noted by Honsell, 'Zur Auskunftsvertrag mit Schutzwirkung für Dritte'; Canaris, C. (1983), pp. 567 et seq.; Kreuzer, K.F. (1976), p. 778; Lorenz, W. (1973), pp. 575 et seq.

517 See Markesinis, B.S. & Uberath, H. (2002), pp. 57-64 and 265-295.

518 See Subsection 2.2 *supra*; on the conditions required by courts for a *VmSzD* to exist, see Markesinis, B.S. & Uberath, H. (2002), pp. 62-64; Beyer, O. (1996), p. 473; Sonnenschein, J. (1989), p. 225; see also BGH *NJW* 1984, pp. 355-356.

choose *VmSzD* over *sgA*.[519] Apart from this, the relevant elements to determine whether a *sgA* exists are all fulfilled, with the exception of the fact that TMOs do not usually issue a warranty on the quality of their trustmarks.[520] Recalling the elements which play a significant role in court decisions on whether or not a *sgA* can be found, first, TMOs know that the information on trustmarks is of great relevance for third-party e-consumers and that it is the basis for important decisions. Second, TMOs use their professional skills in order to issue trustmarks. Third, TMOs have financial interests in providing trustmarks; in fact they are paid for it. Fourth, TMOs do not provide, however, any warranty on the quality of the information provided on their trustmarks; they try to waive any responsibility for the accurateness of the information.[521] Fifth, it remains to be checked, case by case, whether the use that e-consumers make of the trustmarks does not exceed the scope for which they are issued.

The principle of *culpa in contrahendo* seems theoretically applicable to the typical TMO case. TMOs can be seen as professionals that – to some extent – tent to induce third-party e-consumers' confidence in the trustworthiness of the information provided. In fact, it could be argued that the scope of TMO services is to stimulate e-consumers to transact with e-merchants relying on the information provided in the trustmarks, hence the principle of *culpa in contrahendo* may apply. As already pointed out, however, in practice only a very few times was third-party professional liability based on *culpa in contrahendo*.[522]

4. Case law potentially applicable to TMOs

The German case law on third-party professional liability for negligent misstatements is abundant yet not very consistent. The only tendency that

519 See Markesinis, B.S. & Uberath, H. (2002), pp. 265-300; Von Bar, C. (1994), p. 111.

520 In Subsection 2.2 *supra* it was explained that 'if: a) the professional knew that the information he provided was of significant relevance for third-party further decisions; b) the professional prepared and issued the information using his professional expertises; c) the professional has got economic interest in providing such information; d) the professional issued a warranty on the quality of the information provided; e) the third party used the information within the scopes for which it was issued by the professional; a *sgA* will most likely be found.'

521 See Chapter 3, Subsection 5.5 *supra*.

522 See Subsection 2.2 *supra*; see also Canaris (1999), p. 220.

can be identified is that courts tend to move towards an expansion of professional contractual (or quasi- contractual) liability towards third parties. By the same metaphor used in the previous sections, it could be said that courts have been manifestly inclined to widen the 'contractual umbrella'.[523] From the methodological point of view, it is important to analyse the specific reasons on which courts based their decisions to impose liability on professionals case by case, focusing attention especially on the purpose of the duty owed by the professionals towards third parties.[524] The following cases that deal with economic loss caused by negligent statements – typically taking the form of some kind of certification – are extracted from the selection provided by Basil Markesinis in his book *The German Law of Tort*.[525] They are presented to point out the criteria that led the Federal Supreme Court to impose third-party liability in such instances.

4.1 Case 1: Federal Supreme Court (sixth civil senate) 12 February 1979[526]

In the present case, the German branch of a bank asked its headquarters to provide information on the creditworthiness of a third party. Such information was to be passed on to a group of potential investors with whom the German branch was in contact. The information provided by the bank turned out to be inaccurate and one of the potential investors sued the bank, asking compensation for the damages that she suffered by relying on it. The case reached the Federal Supreme Court, which stated that the defendant was liable for breach of its contractual obligations towards the plaintiff, confirming the decision of the lower courts. The Federal Supreme Court stressed that the bank knew that the information which was provided would have been showed to potential investors. Actually, that was the very scope of collecting the information and drafting the related notice. On this point the Federal Supreme Court specified that the information-notice provided by the bank was manifestly prepared to appeal to private individuals (i.e., poten-

---

[523] See Subsection 2.2 and Section 3 *supra*.

[524] This is the method of analysis also suggested by Peter Schlechtriem in his paper *Schutzpflichten und geschützte Personen*. See Schlechtriem, P. (1999), p. 529.

[525] Markesinis, B.S. & Uberath, H. (2002), pp. 265 et seq.

[526] Bundesgerichtshof (sixth civil senate) 12 February 1979, *WM* 1979, p. 548 = *NJW* 1979, p. 1595.

tial investors). Furthermore, the Federal Supreme Court agreed with the position held by the Court of Appeal about the fact that it was clear the defendant was aware that the information was of significant relevance for the plaintiff's further decisions. Moreover, the information provided by the bank was manifestly inaccurate as the defendant omitted facts that were definitely important for a potential investor. On the basis of the assessment made, the Federal Supreme Court then operated a contractual 'assumption'. The Court maintained that, in the case at stake, all the necessary conditions which lead to liability for negligent misstatements were fulfilled.[527] More precisely, given the scope of the information provided, the bank should have realised that the potential investor (as one of the persons to be expected to rely on it) would perceive it as a legally binding statement. Therefore, explained the Court, in this sort of cases, if one of the potential recipients of the information relies on it in good faith in order take further decisions, a contractual relationship arises. Following this line of reasoning, the Federal Supreme Court found a *sgA* to exist between the defendant and the plaintiff and thus the former was under an obligation to provide accurate information to the latter.[528] The bank provided inaccurate information; therefore, it was held liable for breach of its contractual obligation towards the plaintiff.

This is a typical case in which the problem of indeterminate liability arises. Courts generally agree on the fact that it is not necessary that the defendant knows the identity of the relying parties. However, the relying parties need to be part of a determinate group of persons. In the present case, the Federal Supreme Court maintained that the persons to whom the bank provided the relevant information could be identified by virtue of their interests and are part of a calculable group. No relevance was given to the fact that the bank did not know the plaintiff.

4.2 Case 2: Federal Supreme Court (fourth civil senate) 2 November 1983[529]

A dealer asked an officially appointed sworn valuer to provide him with his expert opinion on the value of some real estate properties and the related

---

[527] See Subsection 2.2 *supra*.
[528] See *supra* Subsection 2.2.2.
[529] *WM* 1984, p. 34 = *NJW* 1984, p. 355.

estimated income. The dealer was in contact with some potential buyers for the mentioned real estate properties. Once he had obtained the opinion from the valuer, the dealer showed it to the potential buyers. One of them bought the real estate properties relying on the information provided in the expert opinion, which eventually turned out to be inaccurate. More precisely, the valuer did not consider that the real estate properties were subject to 'social housing' restrictions. It is worth pointing out, however, that the valuer placed in his expert opinion a specific liability disclaimer on this particular issue. The following was stated on this point: 'I do not know whether the premises were put up by private finance or whether they represent social housing. It is not the task of a sworn expert to consider this question. I took great care in ascertaining the rental values in the neighbourhood (…).' Coming back to the buyer, he sued the valuer claiming for the damages suffered by relying on the valuer's expert opinion. Contextually, the buyer maintained that if he had known the correct annual income for the relevant real estate properties, he would not have bought them. In other words, he stressed the direct causal connection between the valuer's mistake and the damages he, the buyer, suffered. Moreover, he insisted that the valuer was fully aware of the fact that the dealer had asked his expert opinion to show it to his potential buyers. Such assertion was not denied by the valuer. Eventually the case reached the Federal Supreme Court after the claim had been denied by both the competent District Court and the Appeal Court. The Federal Supreme Court's decision reversed the previous decisions and allowed the plaintiff's claim. The crucial point in the Federal Supreme Court's reasoning was that the absence of a *VzD* between the plaintiff and the defendant was not a reason enough to exclude defendant's liability. In fact, the Court of Appeal had declared that the dealer's request to provide an expert opinion was not to be considered an offer to the valuer to enter a contract for the benefit of third parties – *VzD*. Hence, the valuer's acceptance to provide the dealer the requested service did not give rise to a *VzD*. On this point, the Federal Supreme Court maintained that in the absence of the substantial conditions for a traditional *VzD* to exist, it was necessary to check whether the plaintiff-buyer was nevertheless under the protective effects of the contractual relationship between the dealer and the valuer (in other words whether a contract with protective effects towards third parties – a *VmSzD* – existed).[530] In deciding that the plaintiff indeed enjoyed the protective

[530] See *supra* Subsection 2.2.1.

effects of the contract, the Federal Supreme Court set forth some general principles on professional third-party liability for negligent misstatements under contract with protective effects to third parties. First, there is no need for the professional to know either who the third parties are or their exact number for the protective effects towards third parties of a contract and the related duties to arise. What is necessary is that the group of third parties to which the duty of care is owed by the professional is capable of being objectively determined. Second, the professional's negligent performance (i.e., the provision of inaccurate information) must have determined the plaintiff's decision which eventually led him to suffer a loss. Third, the Federal Supreme Court, taking into consideration a potential plaintiff's contributory negligence, stated that in the present case the plaintiff, as well as the dealer, did not have the necessary technical knowledge to assess the risk related to the purchase of the real estate properties. This was the reason the support of the valuer was needed. Thus, the plaintiff in this case could not be contributory liable for not having recognised the inaccuracy of the valuer's expert opinion. The Federal Supreme Court went further with its reasoning, stating that, more generally, every contract for the provision of information or advice is based on the assumption that the professional, who provides the information, possesses more knowledge and insight on the matter at stake than the persons who receive it.[531] Thus, if the recipients do not recognise that the information provided by the professional is wrong, this will usually not lead to contributory negligence; and in no instance would contributory negligence of this kind exclude third-party professional liability altogether.

In conclusion, the Federal Supreme Court established that third-party claims for damage caused by the reliance on inaccurate information provided by a professional may be possible also in absence of the necessary conditions for a 'traditional' contract for the benefit of third parties – *VzD*. The Court in fact used its variant: the contract with protective effects towards third parties – *VmSzD*.

4.3 Case 3: Federal Supreme Court (third civil senate) 10 November 1994[532]

This case is particularly interesting because the Federal Supreme Court decided in favour of third-party liability for negligent misstatement even if

[531] See the information asymmetry argument in Chapter 2, Section 4 *supra*.
[532] *BGHZ* 127, p. 378 = *NJW* 1995, p. 392.

the professional (i.e., a surveyor) provided inaccurate information (i.e., a valuation of a real estate) because he received misleading information from his contractual party (i.e., the real estate owner). As to the facts, the owner of a piece of real estate asked a surveyor to issue a valuation of his property. The owner, who was in contact with a potential buyer for his real estate, provided the surveyor with wrong information to obtain a valuation higher than the real value of his property. Relying on the information received by the owner, the surveyor issued an inaccurate valuation. Eventually, the potential buyer purchased the real estate for a much higher price than the actual value of the property because he relied on the information supplied by the surveyor. Once the buyer realised that he had overpaid for the real estate, he immediately sued directly the surveyor and asked him for damages, since the vendor contractually disclaimed any kind of liability towards the buyer for visible or hidden defect of the real estate.

The case reached the Federal Supreme Court, which allowed the third-party action towards the surveyor, overruling the Court of Appeal's decision. The Federal Supreme Court reached such conclusion by interpreting the contract between the surveyor and the owner of the real estate as one having protective effects towards the third-party purchaser – *VmSzD*. More precisely, the Court took the view that the vendor had a factual interest in including the potential buyer in the protective scope of the agreement with the surveyor.

The argument put forward by the Federal Supreme Court was, in first place, that the vendor's and potential purchaser's interests were opposite. The former aimed to have a high valuation of his real estate, the latter a low one. Hence, it was difficult to see how the contractual intention of the vendor was to benefit the third party. However, the Federal Supreme Court stated that in cases of third-party liability for negligent misstatements, such conflict of interests is not relevant to the formation of a contract with protective effects for third parties.[533]

Another criticism was moved against the Federal Supreme Court's line of reasoning. The Court was accused of imposing on the surveyor liability towards the third-party potential purchaser, more than deriving it from the surveyor's contractual will and related explicit or implicit obligations. More precisely, it was pointed out that it was not very likely that the surveyor

[533] See also BGH *NJW* 1987, pp. 1758-9.

would have agreed to assume a duty towards the third party, had it been discussed before entering into the contract with the vendor. However, the Court argued that the surveyor implicitly accepted such duty because he knew that the information he provided was to be relied upon by the third party in order to make an important decision. Furthermore, it is very interesting that the Court stressed the importance of the 'trust' which the potential buyer placed on the surveyor and thus on the information he supplied. The Federal Supreme Court argued that the fact that the potential buyer would most likely give greater weight to the information provided by the surveyor, than to the one supplied by the vendor, justifies the protection for the third party's trust, especially when a malicious vendor tries to conceal the defect of the real estate for sale. Accordingly, the fact that the vendor acted against good faith by providing the surveyor with misleading information on the real estate to value had no bearing on the inclusion of the potential buyer in the protective effects of the contract and the related surveyor's duty to provide the potential buyer with reasonably correct information.

The Court specified further that it was clear that the plaintiff would not have purchased the real estate if the surveyor had issued an accurate valuation, establishing in this way the causal link between the defendant's negligence and the plaintiff's loss. Once again, in this sort of cases, it was considered irrelevant that the defendant did not know to whom the information was to be submitted, being enough for his duty of care to arise that he was aware of the fact that his valuation was intended for a potential buyer.[534]

For all these reasons the Federal Supreme Court decided that a contract with protective effects towards the plaintiff – *VmSzD* – was established and the defendant was liable for breach of his related duties. The Court specified further that the interpretation given to the facts would not expose professionals who provide information to third parties to an excessive liability risk. In fact, if professionals had to provide information without having the chance to verify relevant facts themselves, they would have to state it clearly in their reports and eventually exclude their liability for the accurateness of the information provided.

---

[534] See also *supra* case 2.

4.4 Case 4: Federal Supreme Court (third civil senate) 2 April 1998[535]

In the present case, the plaintiff bought a significant amount of shares of a company, relying on the company's annual audits issued by the defendants-auditors pursuant to Section 316 ff. *Handelsgesetzbuch* (henceforth: HGB). The audits reported inaccurate information, reporting the company's share price as higher than the actual value. Consequently, the plaintiff suffered an economic loss. The Federal Supreme Court allowed the action brought by the plaintiff against the auditors.

In fact, the Court applied also to auditors who carry out the compulsory audit of a company, the familiar principles already described in the cases previously represented in order to establish third-party liability for the provision of negligent misstatements. Firstly, the Federal Supreme Court pointed out that the case law has accepted that protective duties towards third parties can originate from a contract in which one party requests an expert opinion from a professional whose knowledge on the matter has been officially recognised (e.g., auditors, publicly appointed experts). More precisely, the duties arise on the professional towards the third parties to whom the opinion will be passed on. Secondly, and as a corollary of the first argument, the Court stressed that the inner scope of the professional opinion is to induce trust in third parties who can rely on the opinion's evidential value. This will compensate for the fact that the party who commissioned the expert opinion and the third party have conflicting interests for the creation of contract with protective effect to third parties.[536] Thirdly, the Federal Supreme Court specified further that the provision of inaccurate information by the professional will disappoint third-party trust and eventually may cause them a loss. No legal reasons have been found to deny an action for negligence misstatement against the professional to third parties who rely on the inaccurate information to their detriment. However, the Federal Supreme Court took into consideration the intention of the legislator to limit the third-party liability risk of auditors who carry out compulsory audits codified in Section 323 HGB. Nevertheless, the Court, arguing that the *ratio* of Section 323 is to avoid the extension of auditors' liability

[535] *BGHZ* 138, p. 257 = *NJW* 1998 = *JZ* 1998, p. 1013.

[536] See also *supra* case 3 in which the Federal Supreme Court used the same argument.

towards third parties who are not part of a determinate group of persons, allowed third-party cause of action in the present case. It is worth pointing out that, in its decision, the Federal Supreme Court also considered that, from the way the audit was drafted, it was possible to infer that the audit was obviously intended for the use of a third party. Moreover, it was also clear that the third party would not have purchased the shares of the company if the real value of the shares was reported in the audit.

4.5 Some remarks on the cases

The analysed case law confirms the preference of the Federal Supreme Court to allow a cause of action in contract law instead of in tort law to third parties who relied on negligent misstatements provided by professionals and suffered loss. Moreover, it clearly emerges that even if the parties do not expressly commit themselves to contracts which will give rise to duties towards third parties, the Court may derive them from assumptions.[537]

4.5.1 *The contract with protective effect towards third persons –* VmSzD *– as the residual means to establish third-party cause of action*

Among the available contractual means, the Federal Supreme Court very often uses the contract with protective effect for third persons. As already explained, this contract is a judicially created variant of the *VzD* by which the Court manages to include parties who are not even considered in the contract relationship under the contractual 'protective umbrella', allowing third-party direct action to recover pure economic loss from the professional who negligently provides inaccurate information.[538] The contract with protective effect for third persons was used by the Federal Supreme Court as a sort of back-up means to establish a contractual ground for the third-party liability of the professional in case the elements to assert the existence of a *VzD* or a *sgA* were missing.[539] In case 2, the Federal Supreme

[537] See the two striking fictions operated by the Federal Supreme Court in case 3.

[538] See Subsection 2.2 *supra*; see also Markesinis, B.S. & Uberath, H. (2002), pp. 62-64; Beyer, O. (1996), p. 473; Sonnenschein, J. (1989), p. 225; see also BGH *NJW* 1984, pp. 355-356.

[539] See *supra* cases 2 and 3.

Court made this point clear by maintaining that the fact that the Court of Appeal denied that the instruction given to the defendant by the dealer of the real estates to supply an expert opinion constitutes a *VmSzD* did not alone exclude any ground for the plaintiff to bring his claim for damages. Consequently, it was necessary to examine whether the plaintiff was included in the area protected by the contract. Generally, the fact that the defendant professional knew both that the plaintiff was considering to rely on the information the professional provided and that such information would be of great significance to the recipient for taking important decisions were pointed out by the Federal Supreme Curt as the necessary conditions for a contract with protective effect for third persons to exist.[540]

The Federal Supreme Court usually deduces the conditions from the facts or by the defendants' actions. In both case 1 and case 4, for example, the Federal Supreme Court assumed that the professionals knew that the information they provided would be shown to third parties and that such information would be crucial for the decisions the third parties would make by interpreting the style and the content of the bank's notice in the first case and of the auditors' letter in the second case as directed to appeal third parties.

In case 2, the Federal Supreme Court opened the 'contractual umbrella' even wider. The fact that the defendant professional knew that the plaintiff was considering to rely on his information was not necessary to allow the plaintiff's action in the case.[541] The Federal Supreme Court stressed that duties of care can also be created in favour of those persons who are not mentioned by name to the other contracting party. However, the damages can be claimed by the plaintiff under the condition that the defendant's misstatement determined the plaintiff's decision.

4.5.2 *The indeterminate liability issue*

Such permissive approach by the Federal Supreme Court to third-party liability for negligent misstatement brings up the issue of possible indeterminate liability. The Court stated on this point that it is not crucial that the professional was unaware that the information provided was to be submit-

[540] See *supra* cases 1 and 3.
[541] See *supra* case 2.

ted to that particular third-party plaintiff. In fact, it is also not necessary that the contracting party knew the correct number of persons to which his information would be relevant for. The only limit to the extension of professional third-party liability for negligent misstatements set by the Federal Supreme Court consists of the fact that the group of persons who may potentially rely on the information must be capable of being objectively determined.[542]

4.5.3 *Third-party contributory negligence*

Another interesting issue emerges from the analysis of the cases. It concerns the possible role played by third-party contributory negligence in such cases. In case 2, the Federal Supreme Court excluded any third-party contributory negligence on the grounds that a contract for the supply of information or advice is based on the assumption that the information provider possesses more knowledge and insight on the matter than the party who receives the information. Consequently, if the latter does not detect mistakes made by the former, this will not normally be regarded as contributory negligence. The *ratio* of this type of contracts is that the subject's lack of the necessary technical knowledge on a given matter brings him to outsource it.[543] Therefore, if the recipients do not recognise that the information provided by the professional is wrong, this will usually not lead to contributory negligence; and in no instance would contributory negligence of this kind exclude third-party professional liability altogether.[544]

4.5.4 *Third-party liability also in case of false information received by the professional*

Third-party liability for negligent misstatement was also found when a professional received false information from the party who commissioned the service (i.e., his contractual counterpart). More precisely, in case 3, a surveyor overvalued a house because he relied on the false information given

---

[542] See *supra* cases 1, 2, and 3.

[543] Especially on complicated matters, there are no big chances for the subject to spot errors and, furthermore, he should not even be requested to check.

[544] See *supra* case 2.

to him by the real estate owner (i.e., the person who commissioned the survey). The third-party purchaser relied on the surveyor's report, which was drafted on the basis of wrong information maliciously provided to the surveyor by the real estate owner, and bought the house at a price much higher than its real value. The Federal Supreme Court stated on this point that the fact that the vendor acted against good faith by providing the surveyor with misleading information on the real estate to value had no bearing on the inclusion of the potential buyer in the protective effects of the contract and the related surveyor's duty to provide the potential buyer with reasonably correct information.

4.5.5 *The relevance of professional – third party trust relationship*

Moreover, the importance that the Federal Supreme Court gave in the cases dealt with to the trust relationship between the professional and the third-party has to be stressed.[545] Usually, the Court looks at the professional as the key subject who fills the information gap of the third party.[546] In other words, the professional should act as a guarantor who – in case 3, for example – has to protect the third party from the dishonesty of the seller who tries to conceal the true condition of the object for sale. According to the Federal Supreme Court, the trust that third parties – within the protective area of a contract – place in the truth and accuracy of the information provided by a professional must be legally protected even when the incorrectness of the information was caused by the professional's contractual party.[547]

In case 4, the trust argument was also used by the Federal Supreme Court to overcome the obstacles posed by a conflict of interests between the client and the third party in order to include the latter in the protective area of the contract by which the client commissioned a compulsory audit to a professional auditor. In fact, the Federal Supreme Court maintained that, as the purpose of the report is to induce trust in third parties and possesses evidentiary value for them, the protective duty arises upon the auditor also in the case in which he provides the compulsory audit of a company, provided that it appears sufficiently clear to him that the audit is to be used with a third party who trusts in his expert knowledge.

---

[545] See *supra* cases 3 and 4.
[546] See the concept of 'information asymmetry' Chapter 2, Section 4 *supra*.
[547] See *supra* cases 3 and 4.

5. Possible influence of the decisions on TMO third-party liability

The fact that it is not possible to assume the existence of a *VzD* or an *sgA* between TMOs and e-consumers does not exclude the possibility for the latter to claim damages from the former.[548] It will then be necessary to check whether the plaintiff can be included in the area protected by the contract between TMOs and e-merchants (in other words, whether a *VmSzD* may exist). In this respect, according to the decisions of the Federal Supreme Court, if TMOs knew both that e-consumers were contemplating to rely on the trustmarks and that the information provided through the trustmarks would have been of great significance to e-consumers for making important decisions, a contract with protective effect for third persons (e.g., e-consumers) may exist.[549] The two requirements mentioned above were inferred by the Federal Supreme Court from the style and the content of a notice issued by a bank[550] and, in another case, of the audit issued by auditors.[551] In fact, the Court interpreted them as directed to appeal to third parties. It is also evident that a trustmark, which is by definition directed to appeal e-consumers, can be enough to include e-consumers in the area protected by the contract between TMOs and e-merchants. The Federal Supreme Court went even further by stating that the defendant's knowledge of the plaintiff's possible reliance on the information issued by the defendant is not decisive for a plaintiff's cause of action to exist.[552] The damages can be claimed by the plaintiff under the condition that the defendant's misstatement determined the plaintiff's decision. Consequently, if an e-con-

---

548 See *supra* Subsection 2.2 *Contract with protective effects towards third-parties* (*Vertrag mit Schutzwirkung zugunsten Dritter* – VmSzD); *supra* Subsection 4.5.1 *The contract with protective effect towards third persons* – VmSzD – *as the residual means to establish third-party cause of action*; see also *supra* case 2. See extensively on the difference between the *VzD* and the *VmSzD*, Larenz, K. (1956), p. 16. See on the condition required by courts for a *VmSzD* to exist, Beyer, O. (1996), p. 473; Sonnenschein, J. (1989), p. 225.

549 See *supra* cases 1 and 3; *supra* Subsection 2.2.1 *Contract with protective effects towards third-parties (Vertrag mit Schutzwirkung zugunsten Dritter* – VmSzD); See also Markesinis, B.S. & Uberath, H. (2002), p. 63.

550 See *supra* case 1.

551 See *supra* case 4.

552 See *supra* case 2.

sumer manages to prove that the presence of the trustmark on an e-merchant website determined his decision to transact with the latter, he may have a cause of action against the TMO that issued the trustmark.[553]

Third-party contributory negligence has been excluded by the Federal Supreme Court.[554] It is obvious that TMOs possess more knowledge and insight than e-consumers on e-merchant security, privacy, or business practice. Consequently, if e-consumers do not detect mistakes in the information provided (i.e., the inaccuracy of trustmarks), this will not normally be regarded as the e-consumers' contributory negligence. Moreover, the Court concluded that if contributory negligence is found, in no circumstances it will be regarded as so serious as to exclude liability of the professional altogether.[555]

It is interesting to observe that the Federal Supreme Court found third-party professional liability for negligent misstatements also when the professional received false information from the person that asked for his professional performance and is related to him by a contractual relationship. The court stated that in this situation, a contract with protective effects towards third parties – *VmSzD* – can exist anyway.[556] This decision of the Federal Supreme Court can have a very relevant impact on the liability of TMOs. Very often, the auditing procedure carried out by TMOs on e-merchant security, privacy, or business practice is based on internal audit, as opposed to external audit.[557] This means that e-merchants who apply for trustmarks are usually requested to self-assess their policies or practices against the TMOs' standards. Therefore, the accuracy of trustmarks is based upon the good faith of e-merchants. Chances are high that e-merchants declare their security, privacy, or business practices to be up to the standards set by TMOs, even if they are not, in order to receive the trustmarks. Actually, this has already happened.[558] However, even if e-merchants provide false information which leads to the issuance of inaccurate trustmarks which e-consumers rely on and suffer loss as a consequence, the latter will have a

---

[553] Confront on the conditions required by courts for a *VmSzD* to exist, Markesinis, B.S. & Uberath, H. (2002), pp. 59-64; Beyer, O. (1996), p. 473; Sonnenschein, J. (1989), p. 225.

[554] See *supra* case 2.

[555] Ibid.

[556] See *supra* case 3.

[557] See Chapter 3, Subsection 5.2 *supra.*

[558] See Chapter 2, Subsection 6.1 *supra.*

cause of action for damage against TMOs based on the rights of e-consumers deriving from the contract with protective effects for third persons which anyway exists between TMOs and malicious e-merchants. A way for TMOs to possibly avoid liability in this situation is to make clear that the information provided on trustmarks is not verified and that they, the TMOs, exclude their liability for the accuracy of the statements.

It is also interesting to observe the way the Federal Supreme Court perceives professional information providers. They are seen as subjects who fill the information gap of third parties. They assume the role of guarantors who, for instance, protect third parties from the dishonesty of sellers who try to conceal the true condition of the object for sale.[559] In this light, TMOs can be seen as professionals providing the necessary information (i.e., trustmarks) to reduce the information asymmetry between e-consumers and e-merchants.[560] They can also be seen as a sort of guarantors who protect e-consumers from the dishonesty of e-merchants who try to conceal the true condition of their services. The Federal Supreme Court stressed that the trust that third parties, who have been included in the protective area of the contracts between TMOs and e-merchants, place in the veracity of the information provided by such professionals must be legally protected – even when the incorrectness of the information was caused by the professionals' contractual parties.[561]

In conclusion, in order to prevent indeterminate professional liability towards third parties, the Federal Supreme Court set some requirements. In absence of these requirements there cannot be liability.[562] Accordingly, it is not necessary that TMOs know the exact number of e-consumers to whom the trustmarks will be relevant. However, the group of e-consumers must be capable of being objectively determined for liability to exist.

6. Parallel TMOs-CSPs: Applicability by analogy of CSPs' third party liability rules to TMOs

A law which sets forth third-party liability for online professionals comparable to TMOs actually exists. In fact, courts could compare, by analogy,

559 See *supra* case 3 and 4.
560 On information asymmetry, see Chapter 2, Subsection 4 *supra*.
561 See *supra* cases 3 and 4.
562 See *supra* cases 1, 2, and 3.

TMOs to CSPs.[563] The third-party liability of CSPs has been set out in Article 6[564] of the Electronic Signatures Directive.[565] Germany implemented Article 6 of the Electronic Signatures Directive in Section 11 of the Law Governing Framework Conditions for Electronic Signatures (Signatures Law, henceforth: *Sag*)[566] as follows:

> Liability
> '(1) If a certification-service provider infringes the requirements under this Law and the statutory ordinance under Section 24, or if his products for qualified electronic signatures or other technical security facilities fail, he shall reimburse a third party for any damage suffered from relying on the data in a qualified certificate or a qualified time stamp or on information given in accordance with Section 5 (1) Sentence 2. Damages shall not be payable if the third party knew, or must have known, that the data was faulty.
> (2) Damages need not be reimbursed if the certification-service provider has incurred no culpability.
> (3) If a qualified certificate restricts the use of the signature code to certain applications by type or extent, damages shall be payable only within the limits of these restrictions.
> (4) The certification-service provider shall be liable for third parties commissioned under Section 4 (5) and when guaranteeing foreign certificates under Section 23 (1) No. 2 as for his own actions. Section 831 (1) Sentence 2 of the German Civil Code shall not apply.'

According to Section 11 (1) of the *Sag*, if a product or a service provided by CSPs fails, CSPs will have to reimburse third parties for any damage suffered from relying on the data in a certificate, unless the third parties knew, or must have known, that the data were faulty. However, CSPs can avoid reimbursing the damage if they prove that their conduct was not culpable. Moreover, CSPs can limit the use of the certificates. Thus, damages shall be payable only within these limits.[567] This law could be applied by anal-

---

[563] For an explanation of the comparison, see Chapter 1, Subsection 3.1 *supra*.

[564] Article 6. Liability. See *supra* n. 32.

[565] Directive of the European Parliament and of the Council of 13 December 1999 on a Community Framework for Electronic Signatures (99/93/EC) (*OJ* L 13, 19 January 2000, p. 12). For an in-depth analysis of CSP liability as set forth in Article 6 of the Electronic Signatures Directive, see Balboni, P. (2004), pp. 211-242.

[566] Bundesgesetzblatt – BGBl. Teil I S. 876 vom 21. Mai 2001.

[567] For an analysis of such provision, see Heydn, T.J. (2005), pp. 253 et seq.

ogy to TMOs. Accordingly, TMOs will have to compensate e-consumers for any damage suffered from relying on the information in the trustmarks, unless e-consumers knew or must have known that the information was wrong. As it has already been pointed out, however, e-consumers usually lack the necessary technical knowledge to spot information errors.[568] TMOs can avoid liability by proving that they act without *culpa*. In conclusion, damages suffered by e-consumers will only be reimbursed if they occurred within the limits of use of the trustmarks (e.g., the e-consumers suffer damage by relying on the trustmark to make decisions related to e-merchant security, privacy, or business practices).

7. Conclusions

The analysis of the German legal system shows that there are neither specific laws nor case law on TMO third-party liability for the provision of inaccurate trustmarks. However, general tort and contract law provisions may apply to the matter. More precisely, it can be concluded that TMO third-party liability lies on the borderline between tort and contract law.[569]

Section 823 (1) BGB is in general not applicable to losses caused by inaccurate information provided in trustmarks issued by TMOs, but there may be an exception in the instance that e-consumers suffer damage for breach of their privacy right. In fact, privacy right can fall under the protection accorded by Section 823 (1) BGB to 'other rights'.[570] Section 823 (2) BGB may only apply if the losses suffered by e-consumers who relied on inaccurate trustmarks are specifically protected by the law. Moreover, it is very improbable that TMOs issue inaccurate trustmarks with the purpose of harming e-consumers relying on them. Therefore, Section 826 BGB is not very likely to apply. However, if the 'intent' requirement will be interpreted

---

568 See *supra* Subsection 4.5.3 *Contributory negligence*.

569 See Van Dam, C. (2006), pp. 75-76; Zweigert K. & Kötz, H. (1996), p. 617; BGH 23.1.1985 in *JZ* 1985, pp. 951 et seq., noted by Honsell, 'Zur Auskunftsvertrag mit Schutzwirkung für Dritte'; Canaris, C. (1983), pp. 567 et seq.; Kreuzer, K.F. (1976), p. 778; Lorenz, W. (1973) pp. 575 et seq.

570 See Bassenge P. et al. (2006), p. 1248. Van Gerven, W. et al. (2000), p. 142. See also Sections 2, 3 and 5 *supra*.

to include also reckless and thoughtless acts,[571] TMOs' negligence in issuing trustmarks may fulfil such requirement. Nevertheless, it has to be borne in mind that the 'intent' requirement is necessary but not sufficient for the application of Section 826 BGB. In addition, the conduct of TMOs must also be considered *contra bonos mores*. On this point it has to be stressed that recently it has been widely accepted that 'acting *contra bonos mores* implies conduct contrary to the existing economic and legal order or the *ordre public*.'[572] In this respect, providing someone with incorrect information is a type of conduct which falls within the scope of Section 826 BGB.[573] Hence, the conduct of TMOs to negligently provide inaccurate information on e-merchant security, privacy or business practice through the issuance of trustmarks may be regarded as an action *contra bonos mores*.

The present analysis also shows that in German law, third-party professional liability seems to be developing more along contractual or quasi-contractual lines than in tort. More precisely, the Federal Supreme Court often uses the contract with protective effect for third parties – *VmSzD*,[574] which is a judicially created variant of the well-known contract for the benefit of third parties – *VzD*.[575]

According to the decisions of the Court, if TMOs know both that e-consumers are contemplating to rely on the trustmarks and that the information provided through the trustmarks will be of great significance to e-consumers for taking important decisions, a contract with protective effect for third persons (e.g., e-consumers) may exist.[576] Moreover, it seems that no contributory negligence, so serious as to exclude TMO liability, can derive from the fact that e-consumers did not detect the inaccurateness of

---

[571] See Van Dam, C. (2006), pp. 71 and 172; Markesinis, B.S. & Uberath, H. (2002), pp. 292 and 889; Zweigert K. & Kötz, H. (1998), p. 603; Larenz, K. & Canaris, C.W. (1994), pp. 452-453.

[572] Van Dam, C. (2006), pp. 401-402; see also Markesinis, B.S. & Uberath, H. (2002), pp. 889-890; Larenz, K. & Canaris, C.W. (1994), p. 448.

[573] See Van Dam, C. (2006), p. 402.

[574] See more generally on the important role played by *VmSzD* in third-party professional liability, Hirte, H. (1996), p. 388.

[575] See, e.g., case 2. See extensively on the difference between the *VzD* and the *VmSzD*, Larenz, K. (1956), p. 16. See on the condition required by courts for a *VmSzD* to exist, Beyer, O. (1996), p. 473; Sonnenschein, J. (1989), p. 225.

[576] See *supra* cases 1 and 3.

the trustmarks. However, in order not to open the liability floodgates, although it is not necessary that the TMO knows the correct number of persons for whom the trustmark will be relevant, the group of e-consumer that can potentially rely on the trustmark must be capable of being objectively determined by the TMO for its third-party liability to exist.[577]

In conclusion, it has emerged from the analysis that the Federal Supreme Court is at least open to the possible extension of protection in order to compensate for pure economic loss,[578] turning to 'audacious' contractual or quasi-contractual constructions, in the absence of legal ground for a cause of action in tort.[579]

Such attitude could also stimulate courts to apply by analogy Section 11 of the *Sag* according to which TMOs would have to compensate e-consumers for any damage suffered from relying on the information in the trustmarks, unless either the trustmarks have been relied upon for purposes that exceed their scopes or e-consumers knew or must have known that the information was wrong.[580]

---

577 See *supra* cases 1, 2, and 3.
578 Fleming, J. (1989), p. 508; Van Dam, C. (2006), pp. 172-173.
579 See, e.g., case 3.
580 See Section 6 *supra*.

Chapter 7
FRANCE

1. Introduction

In the French legal system, there are no specific provisions on TMOs. Furthermore, no specific cases on TMO third-party liability have been reported. Moreover, literature on TMO third-party liability has not been found either. In this respect France does not differ from England and Germany. Given this scenario, the present chapter aims to investigate whether TMOs' third-party liability for the provision of inaccurate trustmarks exists in France. Approached from another angle, the same question may also be formulated as whether an e-consumer who relies on a trustmark and suffers loss from such reliance can ask the TMO for damages and on which legal ground.

An excursus on the general principles of third-party liability of professionals who provide information will be offered in Section 2. In Section 3, Articles 1382 and 1383 *Code civil*, the two pillars of French tort law, will be analysed. An in-depth analysis, based on literature and case law, on the issues related to 'causation' (one of the three necessary elements of a claim in tort) will be carried out in Section 4. In order to better define the content of information provider obligations, other relevant cases will be dealt with in Section 5. In Section 6, CSP third-party liability rules will be presented. Possible TMO third-party liability scenarios will be showed and discussed in Section 7. Section 8 contains the conclusions.

2. On third-party liability for the provision of information

Generally speaking, under French law, professionals who provide inaccurate information may be liable towards both the parties to whom they are contractually bound and parties outside any contractual relationship, i.e., third parties.[581] Such third-party liability can only be based in tort law. In fact, contrary to what has been shown in the analysis of the German legal system,[582] contractual liability under French law does not expand beyond the parties of the agreement.[583] Furthermore, it is interesting to observe that the breach of contractual duties or obligations is often automatically considered to be a fault in tort. In a leading case, the *Cour de Cassation* recently maintained that if the non-performance of a contractual obligation caused damage to third parties, they can claim compensation in tort. The only thing they have to prove is the causal link between the contractual non-performance and the damage suffered.[584] Moreover, contractually established liability disclaimers do not have effects towards third parties.[585] In other words, every breach of contract may be claimed in tort by aggrieved third parties, regardless of limitations or exclusions of third-party liability set forth by contracting parties.

Tort liability is based on three legal concepts set out in Articles 1382[586] and 1383[587] *Code civil*: 'fault' (*fait fautif*),[588] 'damage' (*dommage*),[589] and

[581] See, e.g., Lemaignan, A. & Nechelis, D. (2005), § 40; Pinna, A. (2003), p. 23.

[582] See Chapter 6, Section 2 *supra*.

[583] See, e.g., Terré, F. et al. (2002), p. 32; Viney, G. (1995), pp. 321 et seq. and 349 et seq.; see also Viney, G. & Jourdain, P. (2001), pp. 447-448. See also Pinna, A. (2003), p. 55.

[584] Principle established in France by Civ. I, 13 February 2001, *Bull. Civ.* I, no. 35; *J.C.P.* 2001.I.338 with obs. Viney, G. (2001) Droit & Patrimoine, no. 2844 with obs. Chauvel, P.: '*les tiers a un contract sont fondés a invoquer tout manquement du débiteur contractual lorsque ce manquement leur a cause un dommage, sans avoir a rapporter d'autre prevue.*' This principle was confirmed in Cass. Ass. plén., 6 October 2006, no. 05-13.255, P+B+R+I, Loubeyre et a. c/ SARL Myr'Ho et a.: *Juris-Data* no. 2006-035298; See also Billiau, M. (2006) 10181; Viney, G. (2006) La responabilité du débiteur à l'égard du tiers auquel il a cause un dommage en manquant à son obligation contractuelle, *Recueil Dalloz 2006 Jurisprudence*, p. 2825.

[585] See Lemaignan, A. & Nechelis, D. (2005), § 48.

[586] Article 1382: 'Any human deed whatsoever which causes harm to another creates an obligation in the person by whose fault it was caused to compensate it.' Article translated in Bell, J. et al. (1998), p. 355.

[587] Article 1383: 'Everyone is liable for the harm which he has caused not only by his deed, but also by his failure to act or his lack of care.' Article translated in Bell J. et al. (1998), p. 355.

[588] See extensively Terré, F. et al. (2002), pp. 693 et seq.; Le Tourneau, P. & Cadiet, L. (2002), pp. 348 et seq.; Viney, G. (1998), pp. 313 et seq.

[589] See extensively Terré, F. et al. (2002), pp. 677 et seq.; Le Tourneau, P. & Cadiet, L. (2002), pp. 1197 et seq.; Viney, G. (1998), pp. 151 et seq.

the 'causal link' between fault and damage (*lien de causalité entre le faute et le dommage*).[590,591] In practice, if the defendant acts in a faulty way, the plaintiff suffers damage, and if a sufficient causal link between these two events exists, the plaintiff will have a cause of action against the defendant. Third-party liability for negligent misstatements does not entail any additional requirement. Therefore, fault, harm, and causation must be proven in order for such claim to be accepted by the court.[592] However, not every act of the defendant will constitute a fault, and not every loss claimed by the plaintiff will be worth compensation. As it will be explained in more detail in Section 4, French courts use the 'causation' requirement as a sort of 'control valve' to allow or stop third-party claims for professional negligent misstatements.[593]

2.1 Obligation of means or obligation of result?

As it has already been pointed out, third parties can claim damages in tort if the non-performance of a contractual obligation is the cause of the damage they suffered.[594] To focus the analysis on tort law, it is relevant to identify the nature of professional contractual obligations in providing information, breach of which can constitute third-party cause of action against the professional.

According to what Clarisse Girot neatly described in her book *User Protection in IT Contracts,* '[c]ontracts of electronic information services are mainly considered contracts of provision of services (*contract d'entreprise*

---

590 See extensively Terré, F. et al. (2002), pp. 815 et seq.; Le Tourneau, P. & Cadiet, L. (2002), pp. 441 et seq.; Viney, G. (1998), pp. 151 et seq.

591 See more generally Van Dam, C. (2006), pp. 46 et seq.; Zweigert K. & Kötz, H. (1998), pp. 615 et seq.; Van Gerven et al. (2000), pp. 57 et seq.

592 See Lyon, 27 January 1994, *Bull. C.N.C.C.* 1994.271 (Annot. P. Merle); Paris, 23 February 1978 Rev. Soc. 79.92; Com., 27 October 1992, Rev. Soc. 79.92; Castell, R. & Pasqualini, F. (1995), p. 87; Guyon, Y. (1994), p. 274; Compagnie Nationale des Commissaires aux Comptes (1995), p. 2; Rennes, 22 February 1999, *Bull. C.N.C.C.* 1999.463 (Annot. P. Merle). On the requirement of causation, see Com., 12 November 1992, *Bull. C.N.C.C.* 1993.89.110 (Annot. M. Jeantin, E. du Pontavice); Com., 15 June 1993, *Bull. C.N.C.C.* 1994.93; Amiens, 20 June 1988, *Bull. C.N.C.C.* 71.316 (Annot. E. du Pontavice); Trib. gr. inst. Cherbourg, 6 April 1976, *B.C.C.* 1976.23.318.

593 Khoury, L. (2001), pp. 457-460.

594 See Section 2 *supra*.

or *de louage d'ouvrage*) in the sense of Article 1710 *Code civil.*'[595,596] She explained further that: 'whereas the information provider, economically speaking, "sells" information, he is, legally speaking, committed to perform a service.'[597] The parallel is easy to draw with auditors (*expert contable*), accountants (*commissaire aux comptes*), surveyors (*agent immobilier*), commercial information offices, financial consultants, etc.[598]

The standard of care that such professionals have to comply with in providing information to their clients is usually decided by courts on a case-by-case basis. In fact, depending on the degree of liability which the courts find appropriate to impose on these professionals, and on the protection that they will grant to their clients, the obligation of professionals may range from an obligation of result to an obligation of means. In the first case, the professional is called to achieve the result expected by his contractual party. In the second instance, the information provider is under the duty to act according to professional diligence and care in providing his services; in other words, he is under an obligation of best effort. Accordingly, if the professional does not achieve the result he will be liable only in the case where he did not act with reasonable care and skill, i.e., if he has been at fault.[599]

The prevalent doctrine and case law tend to see information providers under an obligation of means. The main argument used to support such thesis consists of the fact that the person who ask the professionals for information is always free not to follow it.[600] For example, in the liability of auditors and accountants, the plaintiff has to prove the professionals' negligent action to establish their liability for the provision of inaccurate

---

595 See Lucas A. (1987), p. 488; Groupements Français des Fournisseurs d'Information en Ligne (1986), p. 34; Linant de Bellefonds, X. & Hollande, A. (1992), p. 178.

596 Girot, C. (2000), p. 197.

597 Id.

598 See Chapter 1, Subsection 3 *supra*.

599 See, especially, Tunc, A (1945), p. 449; Pinna, A. (2003), p. 61; and generally Terré, F. et al. (2002), pp. 557-558; Zweigert K. & Kötz, H. (1998), pp. 501-502; Demogue, R. (1928), pp. 1237 et seq., using already the expression: *obligation de moyens et de résultat.*

600 See, e.g., Viney, G. (1975), p. 2750; Cass com. 2 June 1987, D. 1987, p. 500; CA Paris, 25 January 1990, D. 1990, I.R., p. 65; Cass. Com. 27 December 1992, *J.C.P.* 1993 I 22026; Com. 30 January 1974, D. 1974, 428 with note Tendler. CA Lyon 27 October 1971, *J.C.P.* 1972.II.17012, with note R. Savatier. CA Rennes, 21 May 1974, Rev. Banque 1974, 848; RTD com. 1974, 566 obs. Cabrillac, Rives-Lange.

information.[601] In a landmark case, the *Cour de cassation* stated, as a general principle, that a bank is under the mere obligation of means in providing information.[602] As to the facts, the bank provided inaccurate information on the solvency and creditworthiness of a company to a potential investor (who also was a client of such bank). The company went bankrupt a few months after the investor bought its shares, relying on the information received from the bank. As the bank's fault could not be proved, the investor did not manage to receive compensation for the loss suffered. In fact, the company looked solvent at the time the information was provided and there were no elements that could lead to foreseeing its future insolvency.

Nevertheless, a minority case law in which wrong information was sufficient for the *Cour de Cassation* to establish the fault of the provider, has to be given account. For example, in cases concerning the information exchanged between two banks, the *Cour de Cassation* ruled that providing wrong information was sufficient to lead to liability.[603]

2.2 'Reliance' may function as the tongue of the scale between obligation of means and obligation of result

From the analysis carried out so far, it can be argued that information providers are under an obligation of means. Nevertheless, the actual or even potential reliance of their contractual parties on the information in order to take further actions can push the obligation of the information service provider towards an obligation of result. Courts may even assume the reliance of a party who does not possess any knowledge regarding the matter on which the information has been provided. In such cases, the standard of care of professionals will be raised, bringing the professional almost under an obligation of result.[604] The concept of reliance is of course relevant to establish the causal link between the professional's fault in providing inac-

---

601 Lemaignan, A. & Nechelis, D. (2005), § 46.

602 Com. 10 October 1988, *Bull. Joly*, 1988, 931, no. 303; *RD bancaire et bourse* 1989, no. 12, 66, obs. Crédot, Gerard.

603 See Com. 9 January 1978, D. 1978 IR 308, obs. Vasseur; Com. 9 June 1980, D. 1981 IR 192 obs. Vasseur; Com. 24 November 1983, *Bull civ.* IV, no. 322; D. 1984 IR 707 obs. Vasseur; *RTD civ.* 1984, 518 note J. Huet.

604 See Girot, C. (2000), pp. 197-198. See also the concept of information asymmetry dealt with in Chapter 2, Section 4 *supra*.

curate information and the loss suffered by the plaintiff who based his decision on such information. However, it is not that French courts always ask for concrete reliance on statements (e.g., financial), the mere disclosure of documents being sufficient to satisfy the 'causation' requirement.[605] In quite a few third-party claims concerning loss suffered from the provision of inaccurate information, 'causation' was established even when the third parties did not even read the information provided by a professional. In a landmark case, for example, the claim of the creditors of a company against an auditor, who inaccurately audited such company, was allowed by the court even if the creditors did not check the relevant auditor's statement. The rational used by the court to hold the auditor liable towards the creditors was that the auditor's negligent performance, which resulted in an inaccurate statement, allowed the company to stay in business, thus misleading creditors' perception of the company's credit standing.[606]

Contrariwise, it must be noted that there have also been cases in which the court, in its decision, took into consideration whether or not the plaintiff relied on the defendant's statement.[607] Moreover, in other cases, courts have refused to allow compensation because the third-party plaintiff did not exercise reasonable care in relying on a statement or certification.[608]

2.3 Seemingly no worries for indeterminate liability for negligent misstatements

The issue of professional liability to an unlimited group of people for an undetermined amount of money which concerns so much courts and authors in England, Germany, and also in the US seems not to be a major issue

[605] See Com., 21 January 1997, *Bull. C.N.C.C.*1997.406 (Annot. P. Merle).

[606] Tribunal de Grande Instance de Le Havre stated on 15 November 1979 BCC.1980.44, aff'd, Rouen, 27 April 1982, *Bull. C.N.C.C.*47.288. See also Khoury, L. (2001), pp. 460-461.

[607] See Trib. gr. inst. Bordeaux, 31 December 1974, *Bull. C.N.C.C.* 1975.180, aff'd, Bordeaux, 9 February 1976, *Bull. C.N.C.C.* 1976.169; Trib. gr. inst. Montauban, 21 March 1985, *B.C.C.* 1985.234; Trib. gr. inst. Rouen, 9 March 1998, *Bull. C.N.C.C.* 1998.401 (Annot. P. Merle); Paris, 14 November 1997, *Bull. C.N.C.C.* 1998.209 (Annot. P. Merle); Com., 21 January 1997, *Bull. C.N.C.C.* 1997.406 (Annot. P. Merle); Paris, 8 September 1999, *Joly.* 1999.1162 (Annot. J.F. Barbiéri).

[608] See Khoury, L. (2001), p. 461.

in the French legal system.[609] In general, considerations on the foreseeability of damage and the defendants' knowledge of the plaintiff or class of plaintiffs to whom the information will be supplied do not seem to play a crucial role in courts' decisions concerning professional third-party liability for negligent misstatements. Furthermore, no big concern for limiting compensation on the basis of fear of indeterminate liability for information providers is reported by the relevant doctrine.[610] Moreover, as already mentioned, whereas limitation and exclusion liability clauses are in principle valid in contractual liability, they are void if their object is to limit or exclude tortuous liability.[611] However, as it will be shown in Section 4, courts very often use 'causation' as a tool to close the floodgates of professional liability and, thus, to prevent indeterminate liability for professionals.[612]

3. Brief analysis of the relevant rules: Articles 1382 and 1383 *Code civil*

Article 1382
'Any act whatever of man, which causes damage to another, obliges the one by whose fault it occurred, to compensate it.'[613]

Article 1383
'Everyone is liable for the damage he causes not only by his intentional act, but also by his negligent conduct or by his imprudence.'[614]

Articles 1382 and 1383 *Code civil* are the two pillars of French tort law.[615] The combination of the two provisions sets forth the principle that the dam-

[609] See, e.g., Van Dam, C. (2006), pp. 171-172; Van Gerven, W. et al. (2000), p. 236; Zweigert K. & Kötz, H. (1998), p. 617; Tancelin, M. (1993), pp. 108 and 126-127. See also floodgates arguments in Chapters 4, 5 and 6 *supra*.

[610] See, e.g., Khoury, L. (2001), pp. 460-461; Pinna, A. (2003), pp. 55-56.

[611] See the leading case, Civ. II 17 February 1955, D. 1956, 17 with note P. Esmein; *J.C.P.* 1955.II.8951 with note R. Rodière.

[612] See, in particular, Subsection 4.2 *infra*.

[613] Translation from www.legifrance.gouv.fr. Available at <http://195.83.177.9/code/liste.phtml?lang=uk&c=22&r=494#art4453>.

[614] Id.

[615] See Van Dam, C. (2006), pp. 46 et seq.; Terré, F. et al. (2002), p. 683; Zweigert K. & Kötz, H. (1998), p. 615 et seq.; Van Gerven et al. (2000), p. 57 et seq.; Viney, G. (1995), pp. 16 et seq.

age caused to a person by an intentional or negligent conduct (i.e., action or omission) must be compensated. This is a very general clause. Unlike in German law, rights and interests protected by this clause are not limited, or even mentioned. In fact, all the rights or interests, provided that they are legitimate (*intérêt légitime juridiquement protégé*), fall within the scope of Articles 1382 and 1383 *Code civil.*[616] For example, the right to life, physical and moral integrity, specific and general rights of personal privacy (*droit à la vie privée*), as well as property rights, are all fully protected against interference and damage of any kind, including pure economic loss. More precisely, as far as the right to personal privacy is concerned, it is now explicitly recognised and protected in Article 9 *Code civil.*[617]

3.1 Damage

The idea of protected rights or interests[618] relates under French law to the notion of damage, which is the first requirement for a successful claim in tort.[619] In order to be compensated under Articles 1382 and 1383 *Code civil,* the damage must exist, be certain, and be personal to the plaintiff.[620] The nature of the damage (e.g., physical harm or property damage) is not relevant, unlike in the English and the German legal systems.[621] More precisely, pure economic loss has never been excluded from the outset or treated separately.[622] It is compensated under the general principle of liability,[623] as long as it exists, it is certain, and it is personal to the plaintiff. Recalling what was analysed in the previous chapters, the French legal system seems

[616] See Van Gerven, W. et al. (2000), p. 57.

[617] 'Everyone has the right to respect for his private life. Without prejudice to compensation for injury suffered, the court may prescribe any measures, such as sequestration, seizure and others, appropriate to prevent or put an end to an invasion of personal privacy; in case of emergency those measures may be provided for by interim order.' Added to the *Code civil* by the *Loi 70-643 tendant à renforcer la garantie des droits individuals des citoyens* of 17 July 1970, *JO*, 19 July 1970, 6751, D 1970.lég.199.

[618] See Section 3 *supra*.

[619] See extensively Terré, F. et al. (2002), pp. 677 et seq.; Le Tourneau, P. & Cadiet, L. (2002), pp. 1197 et seq.; Viney, G. (1998), pp. 151 et seq.

[620] See Van Gerven et al. (2000), Chapter IV, Subsection 4.1.3 and Chapter VIII.

[621] See Zweigert K. & Kötz, H. (1998), p. 617.

[622] See Tancelin, M. (1993), pp. 126-127.

[623] See Id., p. 108.

to be definitely more open to claims for pure economic loss than the English and the German systems. However, '[t]his does not mean that 'anything goes' in French law.'[624] In fact, French law also has limits to compensation of pure economic loss. As already pointed out, the general rule is that only the breach of legitimate interests enjoy the protection stated by Articles 1382 and 1383 *Code civil* and only insofar as the damage occurred is the direct and certain consequence of the negligent act. Furthermore, courts have significant room for discretion to reject or allow claims. Their decisions are usually centred on the question of whether enough causal link can be established between the act of the defendant and the loss suffered by the plaintiff.[625] Many claims are feasible in theory; nevertheless, a significant part of them are in fact rejected for absence of sufficient direct causation or for lack of certainty of the damage. As already mentioned, the 'causation' requirement is often used by courts to close the floodgates of professional liability in order to prevent indeterminate duties and obligations. Thus, whenever a claim is dismissed in court on the grounds of insufficient causation, the genuine reason might well be the policy considerations, i.e., the defendant should not be burdened with limitless and indeterminate liability.

3.2 Fault

'Fault': a defendant's wrongful act is the second necessary requirement to establish liability in tort. Although there is no official definition of 'fault' provided in the *Code civil*,[626] the literature generally describes it as a behaviour (an action or an omission) which does not conform to relevant rules or standards.[627] More precisely, French scholars conventionally distinguish between *faute délictuelle* and *faute quasi-délictuelle*. The first type

---

624 Van Boom et al. (2004), p. 9.

625 See Trib. gr. inst. Nanterre, 22 Oct 1975, *Gaz. Pal.* 1976.1re sem. Jur. 392 (Annot. A. Plancqueel); Cass. Civ., 11 July 1963, *Gaz. Pal.* 1963. 2e sem. Jur. 389; Cass. Civ. 2e, 19 March 1980, *J.C.P.* 1980.IV.216, D. 1980.IR.414 (Annot. Laroumet).

626 See Le Tourneau, P. & Cadiet, L. (2002), p. 348; Terré, F. et al. (2002), p. 693; Viney, G. (1998), pp. 315-316.

627 See Van Dam, C. (2006), pp. 192-193; Zweigert K. & Kötz, H. (1998), pp. 619 et seq. See also extensively Terré, F. et al. (2002), pp. 693 et seq.; Le Tourneau, P. & Cadiet, L. (2002), pp. 348 et seq.; Viney, G. (1998), pp. 313 et seq.

of 'fault' is the result of an action or omission deliberately put in place by the wrongdoer in order to cause damage. The *faute quasi-délictuelle* consists in a behaviour which produces damage because the wrongdoer did not act with the care expected from a reasonable person in comparable circumstances.[628]

3.3 Causal link between fault and damage

The third necessary element to establish tortious liability under French law is 'causation'. In a nutshell, a link must be found between the defendant's faulty act or omission and the damage suffered by the plaintiff.[629] Given its relevance, an in-depth analysis of the 'causal link' requirement will be separately carried out in Section 4. What is relevant to point out from the very outset is that according to French law, the damage suffered by the plaintiff must be a direct consequence (*un suite immediate et directe*) of the defendant's conduct. No external factors should interfere with such cause-effect chain (*causalité étrangère*).[630]

3.4 Burden of proof

Fault, harm, and causation must thus generally be proven in order to substantiate a claim in tort. Moreover, Articles 1382 and 1383 do not impose any *a priori* limitations as to the specific rights, interests, or group of persons that enjoy protection.[631] Therefore, any plaintiff who is able to prove to have suffered damage, the faulty conduct of the defendant, and the causation link between them can claim compensation under Articles 1382 and 1383.

---

[628] For more on the distinction between *faute délictuelle* and *faute quasi-délictuelle* see Mazeau, H. et al. (1965), numbers 380 et seq.

[629] See Van Dam, C. (2006), pp. 278-279 and 283-284. See also extensively Terré, F. et al. (2002), pp. 815 et seq.; Le Tourneau, P. & Cadiet, L. (2002), pp. 441 et seq.; Viney, G. (1998), pp. 151 et seq.

[630] See Zweigert K. & Kötz, H. (1998), p. 621.

[631] In this respect French tort law is quite different from English and German law. See Van Gerven et al. (2000), p. 58; see also for comparison Chapter 5, Section 4 and Chapter 6, Section 2 *supra*.

4. Focus on 'causation': the major obstacle to claims in tort

Causation as the direct link between the defendant's fault and the damage suffered by the plaintiff can be quite difficult to demonstrate in case of loss suffered relying on inaccurate information. In fact, in this specific instance there are usually a number of external factors that might as well have led to the damage.

Several theories on causations have been developed in the French legal system, two of which are relevant to the present analysis: the theory of adequate causation (*causalité adéquate*)[632] and the equivalence theory (*équivalence des conditions*).[633,634] The first one tries to isolate the direct cause of the damage from the so-called mere circumstances. Indeed, under the adequate causation theory, only the event 'which in the normal state of affairs (*dans le cours habituel des choses*) is of a nature to cause the damage'[635] is relevant in order to establish the causation link in a tortious liability claim. Such theory has been widely accepted by the courts. However, it presents the shortcoming of taking into consideration only one cause even in cases in which several factors contributed to the damage. This issue is overcome by the equivalence theory, which considers all the factors that contributed to the damage. The equivalence theory often plays a significant role in third-party liability cases of auditors.[636] The reason is that the fault of auditors is seldom the only cause of the loss. For instance, it is not unusual that the administrators of companies provide wrong information on which auditors base their accounts. Hence, it can be argued that the damage caused by the auditors' inaccurate statements may originate not only by the auditors' negligence but also by the administrators' lack of care, if not fraud. Finally, above any theory stands the judges' discretionary decisional power. Judges' subjective appreciation of the facts at hand may often prevail on the strict application of legal theories. On the one hand, this enables the deci-

[632] See, e.g., Le Tourneau, P. & Cadiet, L. (2002), pp. 447-448; Viney, G. (1998), p. 161.

[633] See Le Tourneau, P. & Cadiet, L. (2002), pp. 446-447; Viney, G. (1998), p. 160.

[634] For analysis and criticism of these theories, see Terré, F. et al. (2002), pp. 816 et seq.; Flour, J. & Aubert J.L. (1997), §§ 155-62; Malaurie, P. & Aynès, L. (1992), p. 50; Starck, B. et al. (1996), pp. 444-449;

[635] See Khoury, L. (2001), p. 451.

[636] See id. p. 452.

sion process to adapt to different circumstances and needs, thus providing it a certain degree of flexibility; on the other hand, such discretionary power of the judiciary creates some ambiguity as to the necessary causation requirements to be met for a successful claim in tort.

4.1 Looking for a causation test

From the study carried out so far, it clearly comes out that the existence of the causal link between defendant fault and plaintiff damage shall be dealt with case by case. Nevertheless, before carrying out an analysis of relevant cases in order to explain the courts' rationale in determining causation, a very basic causation test will be presented. In fact, given the defendant's fault, the court will usually proceed with the question of whether the loss could have been avoided if the information provider (e.g., an auditor) had acted with the diligence expected of a professional in his position. In case of a negative answer, i.e., that the loss would have occurred anyway, no causal link will be established and the plaintiff claim barred. This is the so-called *conditio sine qua non* test.[637] However, if the answer to the said question is positive, i.e., that the loss would possibly not have occurred, the court will proceed in the causation assessment with a further question: 'Did the inaccurate information provided by the defendant really influence the plaintiff's decision/ action that led to the damage in the specific case?' Causation will generally be found only if the answer to such question is positive.

4.2 Limiting professional liability through causation

The French relevant case law shows that claims against auditors for damage suffered by relying on their statements were quite often rejected when evidence was found that – despite the inaccurate information provided by auditors – the plaintiffs knew, or should have known, about the actual financial situation of the audited companies. Moreover, claims were also not allowed in cases where, even if the auditors' statements presented some mistakes, such mistakes did not impede an understanding of the actual fi-

[637] See, e.g., Viney, G. (1998), p. 169.

nancial situation of the companies.[638] In this respect, the plaintiff's knowledge of the matter at stake is found to be a crucial element in the court's decision. The plaintiff's expertise will be considered by the court in establishing the standard of care against which his conduct should be evaluated. In fact, the court would evaluate not only if the defendant acted with reasonable care to provide his service but also if the plaintiff exercised reasonable care in relying on the statement. The court can actually reject the plaintiff's action for damage if his fault can be established or the court may not allow full compensation.[639] For example, compensation was refused to experts in risk investment who proposed an action for damage against the auditors who failed to capture in their statements the weak financial situation of a company. The court maintained that, given the plaintiffs' expertise on the matter, they should have verified further before making the investment in the company that not much later went bankrupt.[640] Similarly, in another case, a plaintiff-investor sued the accountant and the auditor that certified the accounts of the company which he decided to rescue by making a significant financial investment. A few months after the investment, the plaintiff discovered significant irregularities in the company's accounts. The court maintained that both the defendants and the plaintiff acted negligently. The defendants did not take enough care in analysing the company's accounts to be certified. The plaintiff acted recklessly because he did not verify the accounts. Furthermore, the plaintiff based his decision on accounts that were prepared six months before the time he made the investment. Eventually, the court decided that only half of the damages suffered by the plaintiff should be compensated by the plaintiffs.[641] Courts often see the auditors' negligent conduct as only one of the causes of a plaintiff's damage. Indeed, it is not rare that courts maintained that the plaintiffs' damage originated mainly, or even solely, from the company administrators' negligent (or even fraudulent conduct), or simply from the fact that the

---

[638] See e.g. Com., 15 June 1993, *Bull. C.N.C.C.*1994.93; Com., 12 November 1992, *Bull. C.N.C.C.* 1993.89.110 (Annot. M. Jeantin, E. du Pontavice).

[639] See Trib. gr. inst. Nanterrre, 12 May 1999, *Bull. C.N.C.C.* 1999.489 (the plaintiff was both an auditor and accountant). Contra Trib. gr. inst. Paris, 22 January 1997, *Joly* 1997.432 (Annot. J.F. Barbiéri).

[640] Trib. gr. inst. Paris, 6 April 1990, unreported.

[641] Court of Appeal of Paris, 1 February 1984.

company went bankrupt.[642] This approach does not limit or exclude the auditors' liability from the outset. However, it requires the plaintiff to clearly prove the causal link between auditors' conduct and the loss.

4.3 Presumption of causation

A decision of the *Cour de Cassation* seems particularly relevant to show a possible approach to causation which is rather different from the one taken by the courts in the cases presented above.[643] As to the facts, an accountant negligently certified a company's account which turned out to be inaccurate. The company's shareholders made a capital increase relying on the certified account and eventually suffered a loss. Consequently they brought an action against the accountant under Article 1382 *Code civil.* The *Cour de Cassation* quashed the decision of the Court of Appeal that had previously dismissed the plaintiffs' action in tort. The Court of Appeal based its decision on the fact that the plaintiffs failed to prove the defendant's fault and the causal link between the alleged fault and the loss they suffered. More precisely, according to the court, no evidence was given that the accountant had the chance to verify all the relevant transactions. According to the *Cour de Cassation,* the rationale used by the Court of Appeal in the analysis of the case was contradictory. As pointed out by Alain Viandier in his note to the case, the Court of Appeal 'denied the existence of fault and causation in spite of having acknowledged earlier that the financial statements were "factually inaccurate".'[644] In fact, the *Cour de Cassation* considered that auditors were under an obligation of means. Therefore, it might well have been possible for the auditors to certify inaccurate accounts without being at fault under Article 1382 *Code civil* (e.g., acting with professional reasonable care, but nevertheless not spotting the mistakes in the accounts). However, in the present case, once the plaintiffs proved the factual inaccuracy of the accounts, the defendant was not able to provide evidence that he acted with reasonable care for a professional in a comparable situation. This was enough for the *Cour de Cassation* to overrule the Court

642 See Guyon, Y. (1994), p. 400.

643 Cass. Comm., 17 October 1984, *JCP* 1985.II.20458, annotated by Viandier. Translation by N. Sims.

644 Id.

of Appeal's decision. It can be argued from a close look at the decision that once the defendant's fault was established (because he did not prove the opposite), the *Cour de Cassation* operated a sort of 'presumption of causation' in order to make the auditor liable.

5. Additional case law potentially applicable to TMOs

In addition to the case law already quoted in the previous sections, a selection of other relevant cases dealing with liability for negligent misstatement of professionals that are comparable to TMOs will be presented hereunder in order to further specify the content of information service provider obligations under French law.

The following cases concern the contractual relationship between information providers and their clients. Nevertheless, they are very meaningful in determining information provider third-party liability. Interestingly, the *Cour de Cassation* recently held, in a couple of leading cases, that third parties can claim damages in tort if the incorrect performance of a contractual obligation of the defendant towards his contractual counterpart is the cause of the damage they suffered (*le tiers à un contrat peut invoquer, sur le fondement de la responsabilité délictuelle, un manquement contractuel dès lors qu'il lui a cause un dommage*), adding that this is the only thing they have to prove.[645] Furthermore, it has to be borne in mind that contractual limitation or exclusion of liability clauses do not have effects towards third parties.[646]

[645] Principle established in France by Civ. I, 13 February 2001, *Bull. Civ.* I, no. 35; *J.C.P.* 2001.I.338 with obs. Viney, G. (2001) Droit & Patrimoine, no. 2844 with obs. Chauvel, P.: '*les tiers a un contract sont fondés a invoquer tout manquement du débiteur contractual lorsque ce manquement leur a cause un dommage, sans avoir a rapporter d'autre prevue.*' This principle was confirmed in Cass. Ass. plén., 6 October 2006, no. 05-13.255, P+B+R+I, Loubeyre et a. c/ SARL Myr'Ho et a.: *Juris-Data* no. 2006-035298; See also Billiau, M. (2006), II 10181; Viney, G. (2006), p. 2825.

[646] See Section 2 *supra*. See also Lemaignan, A. & Nechelis, D. (2005), § 48.

5.1 Duty to exercise reasonable care in the collection, processing, and provision of information

Some decisions concerning the liability of commercial information offices that provide specialised information to their clients for a fee are particularly relevant in order to clarify the scope of information provider duties in providing their services. A commercial information office, for example, was found in breach of the duty of care, and thus liable, towards one of its customers for providing him inaccurate financial information on a third party. In fact, the customer asked the office whether it would be reasonable to allow an overdraft to a specific subject. The commercial information office gave a positive answer to its customer who acted accordingly and eventually suffered damage. The customer sued the commercial information office for damage caused by the negligent provision of inaccurate information. The court allowed the plaintiff's action, stressing that the office acted carelessly because, according to the available information, it should have expressed some major doubts on the reasonableness of granting the overdraft.[647] In another case, a commercial information office was found in breach of its duty of care towards a customer because the office provided him only the favourable information on the credit rating of a specific company. For this omission, the commercial information office was held liable of the loss occurred to his customer who relied on this partial information.[648] It is relevant for the present analysis to point out that, as far as commercial information offices are concerned, it is well established in the French case law that they are under an obligation to provide clients with relevant information on which clients may draw their conclusions. In other words, they are under an obligation of means. Furthermore, the service offered by such offices has to be up to their professional standard of care, regardless of the price to which the service is offered. Nevertheless, the price paid by the clients for the information would be relevant in the determination of the amount of damages to be compensated by the commercial information office in case it is held liable for the loss one of its clients

[647] CA Lyon, 27 October 1971, D. 1972, Jur., p. 327, note Tendler.

[648] Cass. Com. 1 March 1982, *Gaz. Pal.* 1982, Som., p. 159; See also Cass. Com. 30 January 1974, *D.* 1974, p. 478, note Tendler; T. com. Nanterre, 7e ch., 10 September 2002: *Juris-Data* no. 2002-190271; CA Paris, 5e ch. C, 28 March 1991: *Juris-Data* no. 1991-021436.

suffered relying on inaccurate information negligently provided by the office.[649]

Moreover, as to the standard of care an information provider has to observe in offering its services, the following case is particularly relevant. A bank was found in breach of its obligation of means towards a client because it provided him wrong information. The key issue in this case was not that the information was wrong. But the court maintained that, first, the bank was aware of the critical relevance for the client of the financial information it was asked for. Second, the bank was held liable for gross negligence because it did not bother to verify the accuracy of the background information that the client passed on to the bank, which actually served as a basis of the relevant financial analysis. Such information turned out to be both wrong and outdated and so the financial analysis supplied by the bank to the client was inaccurate, causing the latter an economic loss.[650] However, even if in the case just described the court took into great consideration the fact that the defendant knew about the relevance of the information to be provided, an auditor in a similar case was held liable under Article 1382 *Code civil* without even having been aware of the fact that his statement would be used by his clients (i.e., company shareholders and also third-party possible investors) as a basis to make decisions. In fact, the auditor failed to mention in his statement the significant passive of the company's accounts.[651]

5.2 Client competence may influence the content of the information provider's obligation; the nature of the information may not

From the analysis carried out so far, it seems quite unchallenged that information service providers are under an obligation of means. In specific circumstances, however, courts may increase information providers' standards of care to a level bringing their obligation close to one of result. This gener-

[649] T.co. Seine, 19 January 1953, *Cimex* v. *Beuzon*, *Gaz. Pal.* 1953, 1, p. 239. See also Girot. C. (2000), p. 198.

[650] Cass. Com 24 November 1983, *R.T.D. Civ.* 1984, p. 518, obs. JH, D. 1984, I.r. 307, note Vasseur.

[651] TGI Grasse, 9 January 1990: *Juris-Data* no. 1990-042186. See also Lemaignan, A. & Nechelis, D. (2005), § 41.

ally happens when information service providers are called to offer their services to persons who do not have much knowledge of the matter on which they asked the information. The reason for such approach by the courts has to be found in the fact that persons who do not possess enough expertise will most likely fully rely on the information provided by the professional.[652] The opposite situation is well represented by the decision of the *Cour de Cassation* in which it was maintained that a bank, in principle, is not even under an obligation to support clients with stock exchange advice, provided that such clients are well informed on the matter. The rule of thumb could be that the information provider's standard of care is inversely related to the expertise of the information recipient. Moreover, it is interesting to note that some doctrines maintained that the information providers shall offer a service which has to be up to the expectations that the providers have created in their clients by offering or describing their services. This means, for example, that if information providers guarantee, either directly or indirectly,[653] to supply reliable information, they may be held liable for breach of contractual duties in case the information is inaccurate. In other words, information providers may be under an obligation to achieve the results they promised to their clients.[654]

As far as the nature of the information is concerned, it is relevant to mention a quite recent case decided by the *Cour de Cassation*. An auditor was sued by shareholders who decided to increase the capital of the company on the basis of a wrongful estimate of future business performance issued by the auditor. The Court of Appeal maintained that there is no liability attached to the wrongful provisions of information on companies' future business performances due to the uncertain nature of this type of information. The court based its decision on the idea that the chances of issuing a wrong estimate are very high, hence the auditor's related liability risk became unbearable. However, the *Cour de Cassation* quashed the decision of the Court of Appeal, stating that, in making its decision, the lower court missed the central issue of the case. In fact, the Court of Appeal fo-

---

[652] See Bertrand, A.R. (1992), p. 51.

[653] Advertising the information service provider as a trustworthy source of information may be enough.

[654] See extensively on this, e.g., Girot, C. (2000), pp. 197 et seq. In the specific instance, the author refers to providers of electronic information.

cused on the type of information provided by the defendant, whereas the crucial question was whether the content of the auditor's estimate influenced the shareholders' decision.[655] From this case, it could be indirectly inferred that the type of information provided by the professional may not have a very relevant impact on his obligations and related liability.

5.3 Occasional qualification of information provider obligation as one of result

A couple of decisions seem quite significant to underlying that, as all the rules occasionally have some exceptions, so does the fact that information provider are generally under an obligation of means. The first case concerns the National Institute of Intellectual Property (INPI), which published wrong information about a politician on its server. Such information originated a scandal that obviously damaged the politician who sued the INPI. The court held the INPI liable under Article 1382 *Code civil*. The fault was not specifically qualified by the court; however, it seemed to consist in the simple provision of wrong information.[656] Similarly, in the second case, a digital server published inaccurate information on the financial situation of a company (i.e., that it was going bankrupt). The company brought a claim against the server, asking for the damages suffered because of the publication of such information. Once again, the court held the information provider liable for the provision of inaccurate information under Article 1382 *Code civil*, without even getting into an analysis of its obligations.[657] Accordingly, it could be argued that, occasionally, the mere provision of inaccurate information is seen as a fault in tort law and thus it leads straight to the related liability.[658] Such conclusion is also supported by some literature. Part of the doctrine holds that professionals that make available to the general public information which is verifiable should guarantee its correctness. In other words, such professionals will thus be exposed to an obligation of result towards relying parties.[659]

---

[655] Cass. 2e civ., 11 December 2003: *Juris-Data* no. 2003-021584.
[656] TGI Bar-le-Duc, 16 January 1997, *D.I.T.* 1998, No. 4, p. 70, note Gavanon.
[657] T. corr. Romans, 14 February 1996, *D.I.T.* 1998, No. 4, p. 74, note Gavanon.
[658] See Girot, C. (2000), p. 201.
[659] See, e.g., Pinna, A (2003), p. 66.

5.4 Some remarks on the analysed cases

The analysis of this additional case law has shown that an information provider is generally under an obligation to collect and process information by adequate means. Furthermore, he has to provide complete (as opposed to partial) information no matter how much he charges and whether he ignores that the information will be used as basis for crucial decisions. Indeed, it is possible to argue that the very purpose of the information provider's activity is to provide its clients with reasonably accurate indications on which they can draw certain conclusions. Moreover, it has been shown that the expectations of the recipients as to the objective quality of the information are likely to be protected under French law insofar they are dependent on such information (e.g., because they have no expertise on the matter), regardless of the nature of the information to be provided. Finally, the mere communication of wrong information has been occasionally considered as a fault in itself under tort law.

6. Parallel TMOs-CSPs: applicability by analogy of CSPs' third party liability rules to TMOs

Courts could also compare TMOs to CSPs and thus apply by analogy the relative liability rules.[660] The third-party liability of CSPs has been set out in Article 6[661] of the Electronic Signatures Directive.[662] France implemented the Directive by Decree 30 March 2001 No. 272. However, in Sections 6 to 9 of the Decree which implemented the provisions on CSPs set forth in the Directive, the extent of the liability of the CSP was not clearly stated. In fact, the liability of CSPs was codified only later in Article 33 of the Law of 21 June 2004 for Confidence in the Digital Economy.[663]

---

[660] See Chapter 1, Subsection 3.1 *supra*.

[661] Article 6. Liability. See *supra* n. 32

[662] Directive of the European Parliament and of the Council of 13 December 1999 on a Community Framework for Electronic Signatures (99/93/EC), *OJ* L 13, 19 January 2000, p. 12. For an in-depth analysis of CSP liability as set forth in Article 6 of the Electronic Signatures Directive, see Balboni, P. (2004), pp. 211-242.

[663] LOI n° 2004-575 du 21 juin 2004 pour la confiance dans l'économie numérique. *J.O* n° 143 du 22 juin 2004 page 11168 texte n° 2.

Article 33
'Sauf à démontrer qu'ils n'ont commis aucune faute intentionnelle ou négligence, les prestataires de services de certification électronique sont responsables du préjudice causé aux personnes qui se sont fiées raisonnablement aux certificats présentés par eux comme qualifiés dans chacun des cas suivants:
1. Les informations contenues dans le certificat, à la date de sa délivrance, étaient inexactes;
2. Les données prescrites pour que le certificat puisse être regardé comme qualifié étaient incomplètes;
3. La délivrance du certificat n'a pas donné lieu à la vérification que le signataire détient la convention privée correspondant à la convention publique de ce certificat;
4. Les prestataires n'ont pas, le cas échéant, fait procéder à l'enregistrement de la révocation du certificat et tenu cette information à la disposition des tiers.
Les prestataires ne sont pas responsables du préjudice causé par un usage du certificat dépassant les limites fixées à son utilisation ou à la valeur des transactions pour lesquelles il peut être utilisé, à condition que ces limites figurent dans le certificat et soient accessibles aux utilisateurs.
Ils doivent justifier d'une garantie financière suffisante, spécialement affectée au paiement des sommes qu'ils pourraient devoir aux personnes s'étant fiées raisonnablement aux certificats qualifiés qu'ils délivrent, ou d'une assurance garantissant les conséquences pécuniaires de leur responsabilité civile professionnelle.'

In a nutshell, this Article creates a presumption of CSP liability for the loss suffered by third parties who reasonably rely on inaccurate information provided in the electronic certificates ('*les prestataires de services de certification électronique sont responsables du préjudice causé aux personnes qui se sont fiées raisonnablement aux certificats présentés par eux comme qualifiés dans chacun des cas suivants : 1. (...),2.,3.,4.*'). However, CSPs can avoid this liability if they prove that they performed the certification service with reasonable care and that they did not voluntarily provide inaccurate information ('*Sauf à démontrer qu'ils n'ont commis aucune faute intentionnelle ou négligence*'). Moreover, CSPs can limit the scope of the certificate, e.g., by stating that the certificate can be used only for certain types of transactions. Accordingly, if third parties use the certificate in transactions that exceed the scope of the certificate, the presumption of CSP liability is not applicable ('*Les prestataires ne sont pas responsables du préjudice causé par un usage du certificat dépassant les limites fixées à son utilisation ou à la valeur des transactions pour lesquelles il peut être utilisé, à condition que ces limites figurent dans le certificat et soient accessibles aux utilisateurs*'). In conclusion, Article 33 requires CSPs to have either a dedicated financial fund for the compensation of the damage suffered by persons who reasonable rely on the certificates or to make an insurance for the same purposes ('*Ils doivent justifier d'une garantie financière suffisante, spécialement affectée au paiement des sommes qu'ils pourraient devoir aux personnes s'étant fiées raisonnablement aux certificats qualifiés qu'ils délivrent, ou d'une assurance garantissant les conséquences pécuniaires de leur responsabilité civile professionnelle*'). It is noticeable that the imposition of a compulsory third-party insurance for CSPs is a major difference between the implementation of the Directive in France. Indeed, in England and Germany, CSPs are not under such obligation.[664]

7. Possible third-party liability for TMOs

On the basis of the analysis carried out, the tortious liability of TMOs towards e-consumers falls under the rules set forth in Articles 1382 and 1383 of the *Code civil*.[665] E-consumers will thus benefit from the general right of action which comes into play whenever TMOs have committed a fault, e-consumers have suffered loss, and there is sufficient causal connection between these two events. Compensation both for pure economic loss and for damage related to the violation of privacy rights may be possible. In principle, there are no specific obstacles for the recovery of economic loss.[666] Once e-consumers manage to prove the TMO's fault, the damage occurred to them, and the causal link between the two, the loss will be compensated. As far as damage related to the violation of privacy rights is concerned, the mere infringement of the right to privacy, protected by Article 9 *Code civil*, suffices to satisfy the conditions of 'fault' and 'damage' for the purposes of

[664] See for comparison Chapter 5, Section 8 and Chapter 6, Section 6 *supra*.

[665] See Sections 2 and 3 *supra*.

[666] See, e.g., Van Dam, C. (2006), pp. 171-172; Van Gerven et al. (2000), p. 236; Zweigert K. & Kötz, H. (1998), p. 617; Tancelin, M. (1993), pp. 108 and 126-127. See also Subsection 2.3 *supra*.

liability under Article 1382 *Code civil*. 'Causation' is the only element that remains to be proved.[667]

7.1 Nature and object of TMO obligations

As a general rule, TMOs may be under an obligation of means: they have to make the best efforts to provide accurate information on e-merchant security, privacy or business practices. Consequently, evidence of negligence is required to hold TMOs liable for the negligent provision of wrongful or deficient information. However, depending on the extent of responsibility which the courts wish to impose on these professionals and on the degree of protection which they will award to the recipients of the information, the classification of TMO obligations can actually vary from an obligation of means to an obligation of result.[668]

Applying by analogy the outcome of the analysis carried out in Section 5 on information providers to TMOs, the latter have to provide their services with reasonable care. More precisely, TMOs are under the obligation to collect and process information by adequate means.[669] Furthermore, they have to provide complete (as opposed to partial) information[670] regardless of the nature of the information,[671] of the fees they charge,[672] and no matter whether they ignore that the information will be used by e-consumers as basis for crucial decisions.[673] The *ratio* behind imposing such obligations consist of the fact that the very purpose of the commercial activity of TMOs

---

[667] See Section 3 *supra*.

[668] See Subsection 2.1 *supra*.

[669] See CA Lyon, 27 October 1971, D. 1972, Jur., p. 327, note Tendler; Cass. Com 24 november 1983, *R.T.D. Civ.* 1984, p. 518, obs. JH, D. 1984, I.r. 307, note Vasseur. See also TGI Grasse, 9 January 1990: *Juris-Data* no. 1990-042186. See also Lemaignan, A. & Nechelis, D. (2005), § 41.

[670] See Cass. Com. 1 March 1982, *Gaz. Pal.* 1982, Som., p. 159; See also Cass. Com. 30 January 1974, *D.* 1974, p. 478, note Tendler; T. com. Nanterre, 7e ch., 10 September 2002: *Juris-Data* no. 2002-190271; CA Paris, 5e ch. C, 28 March 1991: *Juris-Data* no. 1991-021436.

[671] See Cass. 2e civ., 11 December 2003: *Juris-Data* no. 2003-021584.

[672] See T.co. Seine, 19 January 1953, *Cimex* v. *Beuzon*, *Gaz. Pal.* 1953, 1, p. 239.

[673] See TGI Grasse, 9 January 1990: *Juris-Data* no. 1990-042186. See also Lemaignan, A. & Nechelis, D. (2005), § 41.

is to provide e-consumers with relevant information from which e-consumers may draw certain conclusions.[674]

As far as the quality of the information provided by TMOs is concerned, as a general rule, under French law, the provider must meet the user's expectations that he has raised in the formulation of his commercial offer, the description of his service, etc.[675] In cases of TMOs' liability, this general rule unveils a paradox often present in the business practices of TMOs. In fact, as already stressed in Chapter 3, on the one hand, TMOs describe their services (which mainly consist of providing information on e-merchants' practices) as very trustworthy; on the other hand, they tend to contractually disclaim any third-party liability related to the provision of wrongful information.[676] In French law, however, exclusion of liability clauses does not have effects towards third parties.[677] It may be assumed that the expectations raised by TMOs in e-consumers are very high as to the quality of the information provided on e-merchants' practices. Furthermore, e-consumers' expectations on the objective quality of the information are likely to be protected under French law because e-consumers may be seen as very dependent on this information. The dependency of e-consumers is due to their ignorance on matters such as security, privacy, or business practices. Ideally, through trustmarks, e-consumers can receive a sort of guarantee from an independent third party (i.e., a TMO) on the quality of e-merchant security, privacy or business practices.[678] Therefore, e-consumers can be expected to rely totally on the information provided by TMOs. As it was pointed out earlier, the appreciation of the chances that the recipient of the information will actually rely on this information to take further actions can push the obligation of the professional towards one of result.[679] The principle seems to be that the more inexperienced the recipient of the information, the more likely it is that his reliance on the information will be taken for granted and thus the related obligations will be strict.[680] Applying the prin-

---

[674] See T.co. Seine, 19 January 1953, *Cimex* v. *Beuzon*, *Gaz. Pal.* 1953, 1, p. 239.
[675] See Subsection 5.2 *supra*.
[676] See Chapter 3, Subsection 5.5 *supra*.
[677] See Lemaignan, A. & Nechelis, D. (2005), § 48.
[678] See Subection 5.2 *supra*.
[679] See Subsection 2.2 *supra*.
[680] See Girot, C. (2000), pp. 197-198.

ciple to the typical TMO third-party liability case,[681] e-consumers are usually very much inexperienced as far as e-merchant security, privacy or business practices are concerned. Therefore, TMOs may be more easily held liable.[682] However, the necessity of the recipient's reliance on the information for the good outcome of an action in damage for negligent misstatements against the information provider is debated in case law. Moreover, part of the doctrine holds that professionals that make available to the general public information which is verifiable should guarantee the correctness of such information. In other words, it seems that the obligation they have towards third parties is one of result.[683] Accordingly, by issuing trustmarks, TMOs make available to all e-consumers information which is verifiable (i.e., e-merchant security, privacy or business practices). Hence, TMO third-party liability may depend on the fulfilment of the obligation to provide accurate trustmarks to e-consumers.

7.2 Causation issues in TMO liability

As already pointed out in Section 4, the causal link between 'fault' and 'damage' is the most difficult requirement to meet in a claim for damage under Articles 1382 and 1383 *Code civil*. The two main causation theories have been reported: the theory of adequate causation[684] and the equivalence theory.[685,686] Courts enjoy a great deal of discretion on which theory to follow and how to apply it to a given case. If courts apply the adequate causation theory, it may be very difficult for the e-consumer to prove the causal link between the TMO's fault (i.e., the provision of negligent information on e-merchant security, privacy or business practices) and the damage suffered. In these cases, in fact, the e-consumer's damage usually flows

---

681 For the definition of the typical TMO third-party liability case, see Chapter 1, Section 2 *supra*.

682 See Bertrand, A.R. (1992), p. 51. See also, Cass. com. 18 April 1997, *D. Aff.* 1997, p. 369.

683 See, e.g., Pinna, A (2003), p. 66.

684 See, e.g., Le Tourneau, P. & Cadiet, L. (2002), pp. 447-448; Viney, G. (1998), p. 161.

685 See Le Tourneau, P. & Cadiet, L. (2002), pp. 446-447; Viney, G. (1998), p. 160.

686 See Section 4. Moreover, for further analysis and criticism of these theories, see Terré, F. et al. (2002), pp. 816 et seq.; Flour, J. & Aubert J.L. (1997), §§ 155-162; Malaurie, P. & Aynès, L. (1992), p. 50; Starck, B. et al. (1996), pp. 444-449.

from many factors (e.g., the TMO's negligence in performing its services, the e-merchant's lack of care or fraud, the e-consumer's own fault, cunning third-party actions to capture and exploit e-consumer's data). However, the adequate causation theory seeks to eliminate the mere circumstances of the damage to isolate only its immediate cause: the event which in normal state of affairs is of a nature to cause the damage. Hence, the TMO's fault may eventually be considered only a 'mere circumstance' as opposed to an 'immediate cause'.[687]

There will definitely be more chances for e-consumers to succeed in meeting the 'causation' requirement if the court adheres to the equivalence theory. Following this theory, all the facts contributing to the damage are to be retained as causal. Accordingly, the TMO's fault should be taken into consideration as well as, for instance, the e-merchant's lack of care or fraud, the e-consumer's own fault, and cunning third-party action to capture and exploit the e-consumer's data. The equivalence theory is very much in use in the area of auditor liability.[688] Therefore, courts may apply it by analogy to the typical TMO third-party liability case.

However, even if the court applies the equivalence theory, the 'causation' requirement is, nevertheless, very difficult to meet. In fact, as for auditors, the TMOs' fault is usually one of omission rather than of positive action. Therefore, to prove the causal link between TMOs' faulty omission and the damage suffered may be quite challenging for aggrieved e-consumers. Moreover, in most cases, TMOs do not themselves assess e-merchant security, privacy or business practice. Instead, internal audit prevails over external audit as the evaluation practice.[689] In other words, e-merchants that apply for trustmarks are most of the time requested to self-assess their policies or practices against the standards of TMOs. Once again, the fault of TMOs can thus be very difficult to prove for the average e-consumer who usually ignores how the services of TMOs work and does not have any

[687] See generally Guyon, Y. (1994), p. 400. The courts have sometimes linked the plaintiff's damage solely to the fault of the administrators or to the collapse of the company, which was said not to be primarily caused by the auditor. This fact, however, present in every case, does not usually in itself prevent the courts from finding a causal link. Yet, such a fact demonstrates the *prima facie* indirectness of the actions of the auditor as a cause of the failure of the business and the necessity of applying strictly the requirement of causation.

[688] See, e.g., Trib. gr. inst. Paris, 22 January 1997, *Joly* 1997.432 (Annot. J.F. Barbiéri).

[689] See Chapter 3, Subsection 5.2 *supra*.

knowledge of security, privacy or business practices' standards. Given this e-consumers' information gap (also defined as information asymmetry)[690] it will then be almost impossible for e-consumers to prove the causal link between TMOs' fault and the damage they suffered.

Exceptionally, courts have considered it sufficient in order to meet the 'causation' requirement that the plaintiff had shown that the information provided by the defendant was factually inaccurate. In other words, this means that the courts, in those cases, created a presumption of the professionals' liability for the loss suffered by persons who act upon the inaccurate information provided by such professionals.[691] If such presumption is an exception in cases of auditor and information providers liability, conversely, it is a codified rule in CSP liability.[692] In fact, Article 33 of the Law of 21 June 2004 for Confidence in Digital Economy creates a presumption of CSP liability for the loss suffered by third parties who reasonable rely on inaccurate information provided in electronic certificates in undertaking transactions within the scope of the certificates. However, CSPs can escape the liability by proving that they provided their services with reasonable care. Hence, courts may discretionally apply by analogy this presumption to TMO liability cases. It is a matter of fact that if they apply it, e-consumers will have by far greater chances to succeed in their actions for damages against TMOs.

It has to be borne in mind, however, that in cases of professional liability for the provision of wrongful information, courts traditionally tend to deny recovery in favour of third parties on ground of causation even when the professionals were at fault – in order to prevent defendants' exposure to limitless and indeterminate liability.[693]

---

690 See Charter 2, Section 4 *supra*.

691 See, e.g.., Com., 18 October 1994, *Joly* 1994.1302 (Annot. J.F. Barbiéri). See also TGI Bar-le-Duc, 16 January 1997, *D.I.T.* 1998, No. 4, p. 70 note Gavanon; T. corr. Romans, 14 February 1996, *D.I.T.* 1998, No. 4, p. 74, note Gavanon; Trib. gr. inst. Paris, 22 January 1997, *Joly* 1997.432 (Annot. J.F. Barbiéri).

692 See Section 6 *supra*.

693 See Subsections 3.1 and 4.2 *supra*. Moreover, the commentators recognise unanimously that the causation requirement is usually very difficult to meet. See Guyon, Y. (1994), p. 399; Compagnie Nationale des Commissaires aux Comptes (1995), p. 24; Lyon, 27 January 1994, *Bull. C.N.C.C.* 2994.271 (Annot P. Merle) 399; Com., 27 October 1992, JCP.14.G.22026, *Bull. C.N.C.C.* II 91.376, *Rev. Soc.*1993.86 (Annot. M. Jeantin). However, confront Cass. Comm., 17 October 1984, JCP 1985.II.20458, annotated by Viandier. Translation by N. Sims; see also Subsection 4.3 *supra*.

8. Conclusions

The central question addressed in this chapter consists of whether TMOs' third-party liability for the provision of accurate trustmarks exists in France; or, in other words, whether an e-consumer who detrimentally relies on a trustmark issued by a TMO and suffers loss, may ask the TMO for damage and on which ground. It could be argued that, theoretically, TMOs can be held liable in tort under Articles 1382 and 1383 *Code civil* to third-party e-consumers who suffer damage related to the violation of their privacy rights or pure economic loss by relying on inaccurate trustmarks.[694] As a general rule, TMOs may be under an obligation of means.[695] On the one side, thus, TMOs have to make the best effort to provide accurate trustmarks on e-merchant security, privacy or business practices. On the other side, e-consumers have to prove TMOs' negligence to hold them liable for the provision of inaccurate trustmarks. In practice, presenting the proof of the causal link between TMOs' fault and e-consumers' damage represents a major obstacle to the good outcome of e-consumers' actions in tort for damages. E-consumers will have almost no chance of success if the court applies the 'adequate causation' theory. On the other hand, they may have few chances under the 'equivalence' theory.[696] However, courts enjoy a great deal of discretion and may remove the 'causation' obstacle, thus enhancing e-consumers' chances to recover their loss.[697] Courts could apply by analogy Article 33 of the Law of 21 June 2004 for Confidence in Digital Economy. In this way, there will be created a presumption of TMO liability for the loss suffered by third parties who reasonably rely on inaccurate trustmarks in undertaking transactions within the scope of the trustmarks. Accordingly, e-consumers would only have to prove the inaccuracy of the trustmark. Conversely, TMOs may avoid their liability by proving that they issued the trustmark with reasonable care.

---

694 See Sections 2, 3, and 7 *supra*.

695 However, depending on the extent of responsibility which the courts wish to impose on these professionals and on the degree of protection which they will award to the recipients of the information, the classification of TMO obligations can actually vary from an obligation of means to an obligation of result. See Subsection 7.1 *supra*.

696 See Section 4 and Subsection 7.2 *supra*.

697 See Subsections 4.3 and 7.2 *supra*.

Such approach to TMO liability, though desirable by e-consumers, will have to face the more traditional approach that courts have taken in cases involving professional liability for the provision of wrongful information: to deny recovery in favour of third parties on ground of causation.[698]

---

698 See Subsections 3.1, 4.2, and 7.2 *supra*. Moreover, the commentators recognise unanimously that the causation requirement is usually very difficult to meet. See Guyon, Y. (1994), p. 399; Compagnie Nationale des Commissaires aux Comptes (1995), p. 24; Lyon, 27 January 1994, *Bull. C.N.C.C.* 2994.271 (Annot P. Merle) 399; Com., 27 October 1992, JCP.14.G.22026, *Bull. C.N.C.C.* II 91.376, *Rev. Soc.*1993.86 (Annot. M. Jeantin). However, confront Cass. Comm., 17 October 1984, JCP 1985.II.20458, annotated by Viandier. Translation by N. Sims; see also Subsection 4.3 *supra*.

Chapter 8
COMPARATIVE CONCLUSIONS

1. Introduction

In this chapter the conclusions of the comparative study will be drawn and the first two fundamental questions of this book will be answered:

a) *What is TMO third-party liability in Europe?*
More precisely: *Are TMOs liable towards e-consumers who detrimentally rely on inaccurate trustmarks and suffer loss?* From a different perspective: *Do e-consumers have a cause of action to recover the damages they incurred by relying on inaccurate trustmarks directly from TMOs? If they do, what will be the legal basis?*

b) *Is the present TMO third-party liability system in Europe adequate?*

As to the structure, an answer to the first question will be provided in Section 2 after reviewing the findings of the comparative analysis that has been carried out so far. The adequacy of the present TMO third-party liability system in Europe will be challenged in Section 3, concluding in Section 4 by answering the second fundamental question in the sense that the actual system is manifestly inadequate.

2. *De facto* absence of third-party liability for TMOs

A first unchallengeable conclusion is that neither statutory provisions nor case law on TMO third-party liability have been found in the legal systems dealt with. Furthermore, the literature on the matter is scarce if not absent. Therefore, the considerations that follow are logical forecasts based both on decisions that courts took in comparable situations and statutory regulations potentially applicable by analogy.

P. Balboni, Trustmarks in E-Commerce

In none of the legal systems dealt with will it be easy for an e-consumer to recover directly from the TMO damages suffered by relying on an inaccurate trustmark displayed on the website of an e-merchant. E-consumers will have to bring charges against TMOs based either on the general principles of tort or contract law or on statutory provisions and case law that may apply by analogy to TMOs. In most of the cases, e-consumers will have to prove:

a) the damage occurred to them;
b) TMO fault in the issuance of the trustmark; and
c) the causal link between TMO fault and the damage occurred.

However, even if e-consumers manage to prove all these, they could still see their claims rejected by means of adverse policy arguments. Indeed, in the analysed legal systems, the courts enjoy a great deal of discretion. As it will be explained below, in the absence of specific provisions, third-party TMO liability will eventually depend most of all on policy arguments.

As a matter of completeness, before focusing on the European legal systems that have been dealt with, the findings derived by the analysis of the American legal system will be briefly reviewed.[699]

2.1 United States

Under US law, according to Section 552 of the Restatement (Second) of Torts – which is the standard that courts use the most in cases concerning third-party liability for negligent misstatements – the recipient of the information (the e-consumer) has to prove that he suffered relevant pecuniary loss.[700]

Furthermore, it has to be proven that the information is false and that the provider (the TMO) failed to exercise reasonable care in obtaining or communicating it. In order then to establish the causal link between the breach of duty and the damage, courts in the US require plaintiff's reliance on the

---

[699] As already explained in Chapter 1, Section 3 and in Chapter 1, Subsection 3.1 *supra*, since the trustmark phenomenon started in the US and it has expanded there way more than in Europe, the American legal system has been also investigated in order to better understand the relevant legal issues.

[700] See Chapter 4, Subsection 4.3 *supra*. See Feinman, J.M. (2003), p. 43.

defendant's statement both in the misrepresentation action and in the negligent action.[701]

Moreover, according to Section 552 of the Restatement (Second) of Torts, the e-consumer will have to prove that he belongs to the group for whom the TMO intended to supply the information or knew that the recipient of the information intended to supply it.[702]

Finally, it has to be pointed out that in the Jaillet line of cases, the courts applied consistently the contractarian approach (or limited liability theory) as to public policy.[703] All the standards were interpreted in a restrictive way so that no third-party actions in negligence against who disseminates false information were allowed. More precisely, a professional's potential unlimited liability to an indeterminate class and the related chilling effect – originated by the imposition of a high duty of care to a large class of people – which may paralyse their business, are the two main policy concerns.[704]

Very slight seem the chances that the courts will regard the trustmark as an 'expert stamp of approval' (a sort of implied warranty on the accuracy of information). In this case, the applicability of the Jaillet rule will be put under discussion and there may be room for third-party action in negligence against TMOs.[705]

---

[701] See Chapter 4, Section 3 *supra*; Feinman, J.M. (2003), p. 33. See also *H. Rosemblum Inc.* v. *Adler*, 461 A.2d 138, 146 (N.J. 1983); *Greycas, Inc.* v. *Proud.*, 826 F2d 1560 (7th Circ. 1987)); *Barrie* v. *V.P. Exterminators, Inc.*, 625 So. 2d 1007 (La. 1993); *Walpert, Smullian & Blumenthal, P.A.* v. *Katz*, 762 A.2d 582 (Md. 2000). Contrariwise, see *Tri-Professional Realty, Inc.* v. *Hillenburg*, 669 N.E.2d 1064 (Ind. Ct. App. 1996). *In Raritan River Steel Co.* v. *Cherry, Bekaert & Holland*, 367 S.E.2d 609 (N.C. 1988); *Bily* v. *Arthur Young & Co*, 834 P.2d 745 (Cal. 1992).

[702] See *supra* Chapter 4, Subsections 4.3 and 6.1 *Daniel* v. *Dow Jones & Co.* and *Gutter* v. *Dow Jones*; See also Feinman, J.M. (2003), p. 43.

[703] See Chapter 4, Subsection 6.2 *supra*. On the contractarian approach, see Chapter 4, Subsection 5.1 *supra* and more specifically, e.g., Epstein, R.A. (1995), pp. 71-90; Huber, P.W. (1988), p. 5; see also Deakin, S. et al. (2003), pp. 224 et seq.

[704] See *supra* Chapter 4, Subsection 6.1 *Jailet* v. *Cashman*; *Daniel* v. *Down Jones & Co.*; *First Equity Corp.* v. *Standard & Poor's Corp*; *Gale* v. *Value Line*; Subsections 6.2 and 6.3 *supra*. Moreover, see Pacini, C. & Sinason, D. (1999), pp. 495-496; the emblematic Cardozo's statement in *Ultramares* v. *Touche*, where he maintained that liability for negligence 'may expose accountants to a liability in an indeterminate amount for an indeterminate time to an indeterminate class.' 174 N.E. 441 at 444 (N.Y. 1931). See also Siliciano, J.A. (1988), p. 1943. See more generally on the issue, Dobbs, D.B. (2000), p. 1282; Atiyah, P.S. (1985), p. 488; Perlman, H.S. (1982), pp. 71-72.

[705] See Chapter 4, Subsection 6.6 *supra*. See also the relational approach Chapter 4, Subsection 5.2 *supra*. More on the relational approach, see, e.g., Feinman J.M., (1995),

2.2 Europe

Turning our attention to Europe now, the study has highlighted the absence of statutory laws and case law on third-party liability of TMOs. Therefore, the general principles and clauses of tort and contract law apply to the matter. In this respect, it has emerged from the analysis of the three selected European legal systems that the strong arguments used to limit auditor, accountant, and surveyor third-party liability in those jurisdictions[706] may heavily impact TMO third-party liability by limiting it very much if not excluding it at all. However, this trend could be counterbalanced by the possible applicability by analogy to TMOs of the rules set forth in each Member State to implement Article 6 of the Electronic Signatures Directive which prescribe not only a fault-based third-party liability for Certification Service Providers (CSPs) but also set out a presumption of CSP liability towards third parties who suffer loss as a result of their reasonable reliance on CSP certificates. Yet CSPs can rebut the liability presumption by proving that they did not act negligently.[707]

2.2.1 *England*

Under English law, the e-consumer will have first to prove that he suffered damage. Second, the e-consumer will have to prove that the TMO owes a duty of care towards him and then that the TMO acted or spoke in such a way as to breach that duty of care.[708] Third, the e-consumer will have to prove that he suffered damage as a consequence of the TMO's breach of

§ 7.3; Macneil, I.R. (1980), pp. 71-117; Feinman, J.M. (2000), pp. 737-748; Feinman, J.M. (1990), pp. 1299-1304; Gottlieb, G. (1983), pp. 567-612; Macneil, I.R. (1983), pp. 340-418.

[706] See, e.g., the so-called floodgates arguments widely used in Europe to limit third-party professional liability for pure economic loss towards an unlimited amount of people for an indeterminate amount.

[707] The parallel between TMOs and CSPs has been drawn in Chapter 1, Subsection 3.1 *supra*.

[708] See Chapter 5, Section 2 *supra*. See also Powell, J.L. & Stewart, R. (2007), p. 13; See also, Walton, C. et al. (2006), p. 11; Rogers, W.V. H. (2006), pp. 74-75; Dugdale, A.M. & Jones, M.A. (2006), p. 383; Lunney, M. & Oliphant, K. (2007), pp. 120-124; Van Dam, C. (2006), pp. 94-95; Giliker, P. & Beckwith, S. (2004) p. 22; Hodgson, J. (2004), pp. 73-74; Deakin, S. et al. (2003), pp. 74-83; Van Gerven, W. et al. (2000), pp. 44-45; Zweigert, K. & Kötz, H. (1998), p. 609.

duty of care and that the loss suffered falls within the scope of the TMO's duty of care.

Professional duty of care towards the plaintiff and the causal link between professional fault and plaintiff damage are two elements very much interrelated. It could be argued that the two requisites will be met if it is shown by the plaintiff that it was foreseeable for a person in the defendant's position to foresee that his carelessness might cause a loss to the plaintiff, or to the class of persons he belongs.[709] However, 'foreseeability' is necessary to establish the causal link, but it is not sufficient. The proximity requirement has also to be proven. 'Proximity' is about the relationship between the parties: being sufficiently proximate, the defendant would know that his failures might directly affect the plaintiff. It plays a significant role in establishing the necessary causal link between the defendant's act and the plaintiff's loss.[710]

There is no simple test to prove proximity,[711] but there are several factors that, if proven, may help to establish proximity. The most relevant are: (a) facts pointing to the recognition of the defendant's responsibility;[712] (b) the defendant's knowledge of the recipient or class;[713] (c) the defendant's skills or special knowledge of the matter;[714] (d) the defendant's scope of information being sufficiently congruent with the one of plaintiff reliance;[715]

---

709 See on foreseeability Chapter 5, Subsection 4.1 *supra*. See also extensively, e.g., Dugdale, A.M. & Jones, M.A. (2006), pp. 391-392; Walton, C. et al. (2006), pp. 30-32; Lunney, M. & Oliphant, K. (2007), pp. 129-138; Cooke, J. & Oughton, D. (2000), pp. 159-160.

710 See on proximity Chapter 5, Subsection 4.2 *supra*. See also Lunney, M. & Oliphant, K. (2007), pp. 142-143; Dugdale, A.M. & Jones, M.A. (2006), pp. 391-392; Cooke, J. & Oughton, D. (2000), pp. 159-160; Hodgin, R. (1999), pp. 57-58.

711 In *Caparo Industries plc* v. *Dickman*, the House of Lords maintained that 'there is no simple formula or touch-stone to which recourse can be had in order to provide in every case a ready answer.' ([1990] 2 AC 628, HL).

712 See *White* v. *Jones* [1995] 2AC 283-284, HL.

713 See Chapter 5, Subsection 4.2 *supra*. It is considered a proximity factor because of its function of linking the defendant and the plaintiff. In fact, defendant knowledge of plaintiff reliance brings the latter within the group class of persons directly affected by the act or omission of the former.

714 See, e.g., *Hedley Byrne & Co Ltd* v. *Heller & Partners Ltd* [1964] AC 502-503, 505, 510, and 538-539 HL; *Henderson* v. *Merret Syndicates Ltd* [1995] 2 AC 180 HL.

715 See, e.g., *Hedley Byrne & Co Ltd* v. *Heller & Partners Ltd* [1964] AC 503 HL; *Western Trust & Savings Ltd* v. *Strutt & Parker* [1999] PNLR 154, CA. Contrariwise, see

and last but not least (e) the plaintiff's reasonable reliance.[716] It will be deemed reasonable if, in the case at hand, other proximity factors will coexist.[717]

The proximity test, however, may be 'lessened' in case the court decides to apply the 'extended Hedley Byrne liability' test.[718] Accordingly, the plaintiff e-consumer will have to first prove that the facts clearly demonstrate that the defendant TMO had actual knowledge of the likelihood of harm to e-consumers if the trustmark service was inadequately performed and thus the information relating to the e-merchant was inaccurate. Second, it should be proved – or it should clearly result from the facts – that TMOs, in light of the aforementioned knowledge, made a conscious decision to provide information on e-merchant security, privacy, or business practices by supplying trustmarks. Third, e-consumers will have to provide evidence of the tightness of the causal link between TMO inadequate performance of the service and the damage suffered by the e-consumer. In order to demonstrate this, e-consumers will have to prove that few (if any) decisions or acts intervened in the sequence of events leading to the damage. Furthermore, the additional evidence that e-consumers were unable to protect themselves from loss and thus the damage was inevitable will be of help.

It will be radically different if the court decides to apply Section 4 of the Electronic Signatures Regulation on third-party liability of CSPs. CSPs,[719] as already explained, provide services which are comparable to the ones offered by TMOs.[720] If this would be the case, TMOs would be liable to e-consumers who reasonably rely on TMOs' trustmarks – and suffer loss as a result of such reliance – for the accuracy of the information contained in the trustmark at the time of issue, notwithstanding that evidence of TMOs' neg-

---

Caparo Industries plc v. *Dickman* [1990] 2 AC 605, HL where the House of Lords maintained that the identification of the specific purposes for which a statement is issued would not only help in establishing plaintiff's probability of reliance but, more importantly, his entitlement to rely on it.

[716] See Witting, C. (2004), pp. 321-322.

[717] See Id., p. 197.

[718] See *supra* Chapter 5, Subsection 5.2.1 *Proximity*; Witting, C. (2004), pp. 222-225. See more generally on the 'Extended Hedley Byrne liability' line of cases, Lunney, M. & Oliphant, K. (2007), pp. 413 et seq.; Witting, C. (2004), pp. 213 et seq.; Deakin, S. et al. (2003), pp. 124 et seq.; Cooke, J. & Oughton, D. (2000), pp. 199-203.

[719] See Chapter 5, Section 7 *supra*.

[720] See Chapter 1, Subsection 3.1 *supra*.

ligence was not provided, unless TMOs prove that they were not negligent. The standard of care of TMOs would thus consist of providing accurate information on e-merchants on their trustmarks. Aggrieved e-consumers who reasonably rely on a TMO's trustmarks would be entitled to ask the TMO for compensation for the TMO's negligent performance, without having to prove the TMO's negligence. TMOs would be held liable unless they prove that they were not negligent. It will then be crucial at this point to establish whether e-consumers' reliance would have been reasonable. This comes down to proximity. In this respect, it could be argued that e-consumers' reliance is then reasonable, as in a typical TMO third-party liability case the proximity factors are all satisfied.[721] Therefore, TMOs could be held liable.

Eventually, policy arguments will play a crucial role in the courts' decisions on whether or not it will be reasonable to impose a duty of care towards e-consumers on TMOs and thus in allowing a cause of action for aggrieved e-consumers against TMOs that negligently provide wrongful information on e-merchants. More precisely, it will most likely depend on the approach that courts will have to floodgates arguments (i.e., arguments to limit third-party professional liability for pure economic loss towards an unlimited amount of people for an indeterminate amount). So far, courts have tended to 'close the floodgates'. In a typical TMO case, courts may thus maintain that TMOs should be in a position where they are able to weigh the costs of taking precautions against the possible size of claims that could be made against them, and TMOs should be able to predict the number of persons that their negligence might affect. As TMOs are neither able to estimate the cost of claims that they could face nor to predict the number of persons that their negligence might effect, courts could conclude that it would not be fair, just, and reasonable to impose a duty of care towards e-consumers and deny them a cause of action.[722]

---

[721] See Chapter 5, Subsection 6.1 *supra*.

[722] See *supra* Chapter 5, Subsections 4.3 and 6.3 *Floodgates arguments*. See also Lunney, M. & Oliphant, K. (2007), pp. 143-144; Powell, J.L. & Stewart, R. (2007), p. 69; Dugdale, A.M. & Jones, M.A. (2006), pp. 392-398 and 571; Witting, C. (2004), pp. 165 et seq.; Giliker, P. & Beckwith, S. (2004), pp. 94-95; Deakin, S. et al. (2003), p. 115; Cooke, J. & Oughton, D. (2000), pp. 160-161. See on the floodgates arguments, Lunney, M. & Oliphant, K. (2007), p. 458; Hodgin, R. (1999), pp. 61-62; Van Dam, C. (2006), p. 170.

However, it cannot be excluded that courts will hold TMOs liable by applying by analogy third-party liability rules for CSPs set forth in Section 4 of the Electronic Signatures Regulation.

2.2.2 *Germany*

In Germany, first, e-consumers will have to prove that they suffered damage relying on the incorrect information provided through the trustmark by the TMO.

Second, consumers will have to provide evidence that the TMO knew both that e-consumers were contemplating to rely on the trustmark and that the information provided through the trustmark could be of great significance to e-consumers in making important decisions. In this way, e-consumers may in fact be included in the area protected by the contract between the TMO and the e-merchant. A contract with protective effect towards third persons (*VmSzD*) may thus be found between the TMO and the e-merchant. Thus e-consumers may have an action against the TMO accordingly.[723]

Third, e-consumers will have to prove the existence of a causal link between TMO failure to provide accurate trustmarks and the damage they suffered. On this point, e-consumers will have to prove that the presence of the trustmark on the e-merchant's website determined their decision to transact with the latter and that they suffered damage from that transaction.

E-consumers may also invoke the application by analogy of the rules on third-party liability of CSPs, which are set forth in Section 11 *Sag.*[724] In this case, e-consumer will only have to prove that they suffered damage because they had relied on the trustmark and that they did not know that the information provided on the trustmarks were wrong.

A further option for e-consumers could be to try asking the TMO for compensation based on an extensive interpretation of Section 826 BGB.

---

[723] See *supra* Chapter 6, Subsection 2.2.1 *Contract with protective effects towards third parties (Vertrag mit Schutzwirkung zugunsten Dritter* – VmSzD); *supra* Subsections 4.2 and Subsection 4.5.1 *The contract with protective effect towards third persons* – VmSzD – *as the residual means to establish third-party cause of action.* See on the condition required by courts for a *VmSzD* to exist, Markesinis, B.S. & Uberath, H. (2002), pp. 59-64; Beyer, O. (1996), p. 473; Sonnenschein, J. (1989), p. 225.

[724] See Chapter 6, Section 6 *supra.* For an explanation of the possible comparison between TMOs and CSPs see Chapter 1, Subsection 3.1 *supra.*

They would then have to prove that the TMO acted recklessly[725] in providing the inaccurate trustmark and that such act of providing inaccurate information is contrary to the existing economic and legal order and thus the TMO acted *contra bonos mores*.[726]

However, it has to be kept in mind that, in order to prevent indeterminate liability towards third parties, German courts do not require that the professional knows the exact number of people to whom the information issued will be relevant. Nevertheless, they tend to limit professional liability to the group of persons which is capable of being objectively determined.[727]

2.2.3 *France*

Under French law, the first requirement for a successful claim in tort (pursuant to Articles 1382 and 1383 *Code civil*) is to prove the damage.[728] More precisely, the plaintiff e-consumer will have to prove that the damage exists, is certain, and is personal to him.[729]

The second necessary requirement will consist of proving the TMOs' fault, i.e., the fact that the TMOs' performance did not conform to the relevant rules and standards.[730]

The third element that e-consumers will have to prove in order to establish tortious liability is the link between the TMOs' faulty act and the damage suffered.[731] More precisely, e-consumers will have to prove that if the

725 See *supra* Chapter 6, Subsection 2.1 *Section 826*; and Section 7 *supra*. See also Van Dam, C. (2006), pp. 71 and 172; Markesinis, B.S. & Uberath, H. (2002), pp. 292 and 889; Zweigert K. & Kötz, H. (1998), p. 603; Larenz, K. & Canaris, C.W. (1994), pp. 452-453.

726 See *supra* Chapter 6, Subsection 2.1 *Section 826*; and Section 7. See also Van Dam, C. (2006), pp. 401-402; see also Markesinis, B.S. & Uberath, H. (2002), pp. 889-890; Larenz, K. & Canaris, C.W. (1994), p. 448.

727 See Chapter 6, Subsections 4.2, 4.3, and Section 5 *supra*.

728 See Chapter 7, Subsection 3.1 *supra*. See extensively Terré, F. et al. (2002), pp. 677 et seq.; Le Tourneau, P. & Cadiet, L. (2002), pp. 1197 et seq.; Viney, G. (1998), pp. 151 et seq.

729 See Chapter 7, Subsection 3.1 *supra*. See also Van Gerven, W. et al. (2000), Chapter IV, Subsection 4.1.3 and Chapter VIII.

730 See Chapter 7, Section 3.3 *supra*. See Van Dam, C. (2006), pp. 192-193; Zweigert K. & Kötz, H. (1998), pp. 619 et seq. See also extensively Terré, F. et al. (2002), pp. 693 et seq.; Le Tourneau, P. & Cadiet, L. (2002), pp. 348 et seq.; Viney, G. (1998), pp. 313 et seq.

731 See Chapter 7, Subsection 3.4 *supra*. See Van Dam, C. (2006), pp. 278-279 and 283-284. See also extensively Terré, F. et al. (2002), pp. 815 et seq.; Le Tourneau, P. & Cadiet, L. (2002), pp. 441 et seq.; Viney, G. (1998), pp. 151 et seq.

TMOs had acted with due diligence the damage would not have happened.[732] Furthermore, evidence will also have to be provided that the inaccurate trustmark had in fact influenced the e-consumers decision that led to the damage. The causation requirement is the most difficult to meet.[733]

E-consumers may also try to argue for the application by analogy of the CSPs' third-party liability rules set forth in Article 33 of the Law of 21 June 2004 for Confidence in Digital Economy. Accordingly, e-consumers will only have to prove the inaccuracy of the trustmark. Conversely, TMOs may avoid their liability by proving that they issued the trustmark with reasonable care.

However, it has to be stressed that in cases of professional liability for the provision of wrongful information, courts traditionally tend to deny recovery in favour of third parties.[734] Many claims are feasible in theory; nevertheless, a significant part of them are rejected for absence of sufficient direct causation. As already mentioned, the 'causation' requirement is often used by courts to close the floodgates of professional liability in order to prevent indeterminate duties and obligations. Thus, whenever a claim is dismissed in court on the grounds of insufficient causation, the genuine reason might well be the policy considerations, i.e., the defendant should not be burdened with limitless and indeterminate liability.[735]

2.2.4 *Conclusions*

At the end of this brief excursus on the various scenarios that e-consumers will face if they want to ask TMOs for compensation for damage suffered for the provision of inaccurate trustmarks, the following conclusions can be drawn.

---

[732] See Chapter 7, Subsection 4.1 *supra*. See on the *conditio sine qua non* theory, e.g., Viney, G. (1998), p. 169.

[733] See Chapter 7, Subsection 7.2 *supra*.

[734] See Chapter 7, Subsections 3.1 and 4.2 *supra*. Moreover, the commentators recognise unanimously that the causation requirement is usually very difficult to meet. See Guyon, Y. (1994), p. 399; Compagnie Nationale des Commissaires aux Comptes (1995), p. 24; Lyon, 27 January 1994, *Bull. C.N.C.C.* 2994.271 (Annot P. Merle) 399; Com., 27 October 1992, JCP.14.G.22026, *Bull. C.N.C.C.* II 91.376, *Rev. Soc.*1993.86 (Annot. M. Jeantin). However, confront Cass. Comm., 17 October 1984, JCP 1985.II.20458, annotated by Viandier. Translation by N. Sims; see also Subsection 4.3.

[735] See Chapter 7, Subsection 3.1, Section 4, Subsection 7.2 *supra*.

Very broad liability clauses interpreted restrictively by means of public policy considerations based on fear of liability in an indeterminate amount for an indeterminate time to an indeterminate class characterise the common law approach to third-party liability for negligent misstatements. In fact, the chances for aggrieved e-consumers to hold TMOs liable for the provision of inaccurate trustmarks are very slim under English law as well as in the US.

The German and the French legal systems offer more legal grounds on which to base third-party liability claims against TMOs. French courts, however, have proved their inclination to use 'causation' to impede or at least limit third-party liability for negligent misstatements; whereas, German courts seem to be more open in that respect.

Moreover, in Europe, the rules on third-party liability of CSPs, derived from the implementation in the Member states' legal systems of Article 6 of the Electronic Signatures Directive, could apply by analogy to TMOs. Therefore, these provisions may offer some additional protection to e-consumers who rely on inaccurate trustmarks, provided that courts decide to take them in consideration.

Within the analysed European legal systems, the French is the one which grants, at least in theory, more protection to third parties who suffer loss by relying on negligent misstatements. The general tort clauses set out in Article 1382 and 1383 *Code civil* applies to the situation at stake. However, it will be quite difficult for e-consumers to prove the causal link between TMOs' fault and the damage. Under German law, the most suitable means for e-consumers to recover their loss from TMOs seems to be represented by the 'contract with protective effect for third parties' (*VmSzD*), which brings strangers to a contract under its 'protective umbrella'. The court, though, requires the group of 'strangers' (i.e., e-consumers) to be capable of being objectively determinate by the tortfeasor (i.e., the TMO). Given the fact that a trustmark can be seen and relied upon by a potentially unlimited number of e-consumers, it is unlikely that the mentioned requirement will be met. As far as English law is concerned, a proximity relation between TMOs and e-consumers, necessary for TMOs' duty of care towards e-consumers to arise, seems to be established under the 'extended Hedley Byrne' liability rules. However, the courts' strong aversion to grant compensation for pure economic loss is very likely to bar any of such third-party claims against TMOs under English law.

In summary, the answer to the first fundamental question addressed in this book: *Are TMOs liable towards e-consumers who detrimentally rely on inaccurate trustmarks and suffer loss?* is that, in theory, there is enough legal ground to enforce TMOs' third party liability; in practice, however, the chances that TMOs will not be liable towards aggrieved third-party e-consumers who relied on the misleading or false information provided in the trustmark are way greater than the chances that TMOs will be held liable. At present, thus, it is possible to conclude for a *de facto* absence of TMOs liability towards e-consumers in Europe.

3. Challenging the adequacy of the present TMO third-party liability system

Unregulated market forces, characterised by the opportunistic behaviour of the market players, are driving TMOs more towards an untrustworthy practice than otherwise. TMO practice seems to be going the wrong direction.[736]

The analysis carried out in Chapter 3 showed that the necessary conditions for a trustworthy certification practice:

1. certifier independency,
2. impartiality in the auditing procedure,
3. active monitoring of the certificate owner's practice,
4. certifier enforcement power, and
5. certifier accountability

are either missing or not fully guaranteed in the services of TMOs. In fact, as it was already pointed out, low-price policy and the widespread use of private sponsorships heavily challenge TMO independence.[737] Moreover, a third-party independent audit of e-merchants who asked for a trustmark is an exception while self-assessment by e-merchants is the rule.[738] Active monitoring of e-merchant compliance and enforcement procedures are gen-

[736] This has been showed in Chapter 3 *supra*.

[737] See the 'low-balling effect' theory in Chapter 3, Subsection 5.1. See Makkawi, B. & Schick, A. (2003), pp. 797-822; Calegari, M. et al. (1998), pp. 255-275; De Angelo, L.E. (1981), pp. 113-127.

[738] See Chapter 3, Subsection 5.2 *supra*.

erally in place. However, due to the lack of transparency and information, it is difficult to assess the effectiveness of the present monitoring activities and enforcement practices. It is also legitimate to doubt, first, whether monitoring activities are in fact carried out and, second, whether enforcement procedures are activated at all.[739] As to TMO accountability, this book has revealed the absence in Europe (as well as in the US) of both specific statutory provisions and case law on TMO third-party liability. Furthermore, although in theory there seems to be enough legal ground to enforce TMO third party liability on the basis of the general principles of tort and contract law, the chances in practice that TMOs will not be liable towards aggrieved third-party e-consumers who relied on the misleading or false information provided in trustmarks are way greater than chances that TMOs will be held liable. Therefore, it is possible to conclude that TMOs are in fact not accountable.[740]

It results in a picture of a rather untrustworthy TMO practice.

None of the stakeholders (i.e., e-consumers, e-merchants, governments, and TMOs) seem to profit from an untrustworthy TMO practice.[741] On the one side, the risk of undetected e-merchants' actions that can cause loss to e-consumers is high and damages are very difficult to recover.[742] On the other side, by joining a trustmark programme, e-merchants run the risk that other e-merchants with the same trustmark will violate the rules of the trustmark programme and eventually cause damage to e-consumers who happen to rely on that trustmark. In this way, the reputation of that trustmark programme will be damaged and the investments made by all e-merchants to join it will be wasted. Although this is a risk that will always exist, an unreliable TMO practice will exponentially increase it. More generally, damage to e-consumers caused by a certified e-merchant will decrease the amount of trust that e-consumers have in e-commerce. Consequently, this will not bring any positive effect to e-business, nor to the economy. Moreover, governments run the risk in letting untrustworthy TMOs run wild. In fact, if damage related to e-consumers' reliance on trustmarks occurs, the inactivity of government in regulating the matter will also be indicated as

[739] See Chapter 3, Subsection 5.3 *supra*.
[740] See *supra* Section 2.
[741] See Chapter 2, Section 8 *supra*.
[742] See *supra* Section 2.

one of the reasons for TMOs' unreliable practice. A sound trustmark practice is a valuable means of self-regulation that gives governments the chance to stay out of a sector of Internet regulation. An untrustworthy trustmark practice may require a prompt corrective government action. Last but not least, none of the TMOs have reached critical mass yet. If e-consumers suffer damage because of their reliance on trustmarks, the soundness and the reliability of the whole trustmark system will be questioned and once e-consumers' trust is lost it will be very difficult to win it back. Consequently, the TMO business will stop even before it has started.

4. Conclusions

Liability law has traditionally been one of the first legal guards against undesirable societal developments.[743] In this respect, liability law aims to discourage subjects from putting in place specific unwanted conducts by making them liable for them and eventually obliging them to compensate the damages that may flow from their misconducts.

The actual TMO third-party liability system – based on the general principles of tort and contract law which result into a *de facto* absence of TMOs liability towards e-consumers – is not adequate because, to use an euphemism, it does not work as a deterrent to discourage TMO untrustworthy practice. In fact, the total lack of specific rules on TMO liability, coupled with the practical impossibility for e-consumers to hold TMOs liable in tort or in contract, as shown in this book, creates a sort of legal 'immunity' for TMOs, which surely has not helped foster best practice.

A strong consolidated argument against an expansion of liability in auditor and accountant literature is that it does not produce a significant incentive to exercise care because accountants already have sufficient incentives to audit with due care.[744] These incentives include malpractice liability to clients, sanction by regulators in security cases or other regulated matters, concern for reputation, and the threats of professional disciplinary proceedings. Such argument, however, is not applicable to TMOs. There is almost no malpractice liability to e-merchants because it is most of the time almost

---

[743] See Schellekens, M. & Prins, C. (2006), p. 49.
[744] See, e.g., Siciliano, J.A. (1998) p. 1953.

fully contractually waved. No sanctions have been imposed on TMOs by regulators so far. Reputation is a key element for the success of TMOs, TMO reputation and credibility being positively related to e-consumer trust in e-merchants.[745] Nevertheless, notwithstanding the scandal reported in the US in the late 1990s,[746] TMO practice does not seem to have improved since then, proving that concern for reputation does not really work as an incentive to exercise care. Another reason for this can be that the chances that inaccurate trustmarks will be detected are not really high, thus the lack of TMOs' concern for losing their reputation.[747] Furthermore, issuing trustmarks is not a regulated profession; thus, TMOs are not subject to any professional disciplinary proceedings.

Moreover, floodgates arguments[748] – commonly used in the US and Europe to limit third-party professional liability for negligent misstatements causing pure economic loss potentially to an unlimited amount of people for an indeterminate amount – have also to be taken into consideration. In this respect, it could be argued that implementing specific rules on TMO liability will not alone lead to a flood of third-party claims possibly resulting in the financial ruin of TMOs. To support what was just stated, the

---

[745] See Chapter 2, Subsection 5.2 *supra*. See Wakefield, R.L. & Whitten, D. (2006), pp. 1-19; Lafferty, B. et al. (2002), pp. 1-12; Hoxmeier, J. (2000), pp. 115-139; Goldsmith, R. et al. (2000), pp. 43-54; Keller, K.L. (1998).

[746] See Chapter 2, Subsection 6.1 *supra*.

[747] Generally, e-consumers do not have enough competence on TMO practices to spot an inaccurate trustmark. See Chapter 2, Section 4 *supra*. See also Nordquist, F. et al. (2002). Moreover, the de-materialisation of information (e.g., personal data, credit card data, etc.) brought about by the Internet (see Chapter 2, Section 3 *supra*) makes it very difficult for e-consumers even to realise that their data have been stolen – e.g., by some cunning third-party who took advantage of the low security of an e-merchant website – or used by e-merchant in an unlawful way or in a way which is incompatible with the purposes stated on the relevant trustmark.

[748] The rationale of the floodgate liability theory was very well summarised by Cardozo C.J. in *Ultramares Corp. v. Touche* who maintained that to allow such recovery would 'expose [defendants] to a liability in an indeterminate amount for an indeterminate time to an indeterminate class' ([1931] 255 N.Y. 179). In other words the meaning of this metaphor is that the defendant – who negligently drafts an inaccurate statement which is relied upon by a potentially unlimited number of people and causes them economic losses – would be flooded with claims, resulting in his financial ruin. On the floodgates arguments see extensively, Lunney, M. & Oliphant, K. (2007), p. 458; Hodgin, R. (1999), pp. 61-62; Van Dam, C. (2006), p. 170; Cane, P. (1996), pp. 455 et seq.

example of the specific CSPs' liability provision set forth by the European legislator in Article 6 of the Electronic Signatures Directive could be mentioned. Such provision has been implemented in all the European legal systems and CSPs are still on the market providing their services. Going down this line of reasoning, one could say that floodgates arguments are not a reason not to lay down specific rules on TMO liability. However, they have to be taken into consideration in the process of drafting such rules, providing TMOs with the possibility of reasonably limiting their liability as it happens, e.g., for CSPs.

Finally, the apathy of European governments on the matter does also not seem to be adequate to the present phase of the Internet. Nowadays, the Internet can be seen as being in the transition between the so-called 'creative anarchy' phase and the phase of 'rules'.[749] In fact, after some years of anarchy and self-regulation, not only e-consumers but also enterprises are calling for government coherent regulatory intervention to make the Internet a safer and neater marketplace, especially when the market does not manage to properly regulate itself.[750]

Hence, it seems reasonable to answer the second fundamental question: *Is the present TMO third-party liability system in Europe adequate?* by concluding that the present TMO third-party liability system is manifestly not adequate. Therefore, the next chapter will be dedicated to the elaboration of a model of adequate third-party liability for TMOs.

---

[749] Debora Spar, in her book Ruling the Waves, distinguishes four phases through which the society that exploits discoveries usually passes. These phases are: 1. innovation; 2. commercialisation; 3. creative anarchy; and 4. rules. The phase of 'innovation' is defined as the flash point of discovery. The phase in which pioneers (or pirates, depending on the perspective) move into the new area seeking to exploit its potential is described as the 'commercialisation' one. 'Creative anarchy' is pictured as the phase when the needs of ordinary commerce come into tension with the freewheeling spirit of the new frontier. The final phase, i.e., 'rules', follows unavoidably as the commercial enterprises unable to hold back anarchy on their own call for government intervention as the best vehicle for bringing order (and profit) to the wild frontier. Spar, D.L. (2001), pp. 1-22. See also, Mann, R.J. & Belzey, S.R. (2005), pp. 239 et seq.

[750] Spar, D.L. (2001), pp. 1-22; Mann, R.J. & Belzey, S.R. (2005), pp. 239 et seq.

Chapter 9
RECOMMENDATIONS: TOWARDS AN ADEQUATE TMO THIRD-PARTY LIABILITY SYSTEM

1. Introduction

In the previous chapter, it was concluded that the TMO liability system is not adequate. Hence, this chapter will be dedicated to answering the third fundamental question of the present analysis, which is: *What will an adequate TMO third-party liability system be?*

In Section 2, the reason of the choice to combine law and ethics in order to build a model of adequate third-party liability for TMOs will be provided. Section 3 will be dedicated to the actual elaboration of the model for adequate third-party liability for TMOs based on the theory of 'Warranted Trust'. Such theoretic model will be compiled in Section 4, providing a concrete proposal for an adequate TMO third-party liability system. In Section 5, the proposed adequate third-party liability is summarised into a liability provision. How the provision would work in practice will be explained in Section 6. Section 7 contains some conclusions and recommendations for international regulatory bodies.

2. Law and ethics: two complementary normative sciences

As already pointed out in Chapter 1, there are two main reasons why an ethical theory is chosen in order to improve the TMO third-party liability system. First, this approach seems to widely comply with what has been recommended at the European level. In fact, in Recital 32 of the Directive 2000/31/EC of the European Parliament and of the Council of 8 June 2000 on Certain Legal Aspects of Information Society Services in the Internal

Market,[751] the creation of a code of conduct at the Community level is encouraged as a means to determine the rules on professional ethics applicable to commercial communication.[752] As the present effort aims to contribute to the creation of European standards/code of conduct for TMOs, the ethical aspects of TMO practice need to be taken into consideration also in setting out the related liability system. Second, law and ethics are two normative sciences which traditionally can very well complement each other. Simply put, ethics sets the basic societal interests that law should guarantee: '[w]ithin the tradition of natural law thinking which finds its roots in the philosophies of Aristotle and Aquinas, the political community has generally been understood in terms of a fundamental goal: that of fostering the ethical good of citizens. Law, on this concept, should seek to inculcate habits of good conduct and should support a social environment which will encourage citizens to pursue worthy goals and to lead valuable lives.'[753]

The core of TMO liability will be based on the same principles of surveyors', auditors', and accountants' liability so that e-consumers will receive at least the same protection in the online world as in the more traditional paper-based and oral forms of communication.[754] Once such basis of liability is created, it will be tailored to the specific TMO practice by developing the concept of 'adequacy' in order to make it more effective.

In this respect, the concept of 'adequacy' will be defined by applying to the trust relationship between TMOs and e-consumers the ethical theory of Warranted Trust – which, in a nutshell, aims to protect trustor reliance on trustee by implementing a regulative framework which takes into consideration the interest of both parties and the influence of the specific context in which the trust relationship develops – and considering the social, eco-

---

[751] *OJ* L 178, 17.7.2000, p. 1.

[752] Recital 32: 'In order to remove barriers to the development of cross-border services within the Community which members of the regulated professions might offer on the Internet, it is necessary that compliance be guaranteed at Community level with professional rules aiming, in particular, to protect consumers or public health; codes of conduct at Community level would be the best means of determining the rules on professional ethics applicable to commercial communication; the drawing-up or, where appropriate, the adaptation of such rules should be encouraged without prejudice to the autonomy of professional bodies and associations.'

[753] Craig, E. (2005), p. 542.

[754] This aim is often referred to as 'what applies offline should apply online'. See Schellekens, M. (2006), pp. 51-76; Koops, B.J. et al. (2000).

nomic, and political value of trustmarks. More specifically, the theory of Warranted Trust should help find a legal solution to the fundamental issue of the so-called expectations gap.[755] In fact, on the one side, TMOs try to waive any possible liability towards e-consumers on the accuracy of the information they provide on e-merchant security, privacy, and business practices through trustmarks.[756] On the other side, e-consumers would trust TMOs and rely on trustmarks as they tend to perceive TMOs as guarantors of e-merchants' practices. Through the theory of Warranted Trust, it will be possible to take into consideration both TMOs' and e-consumers' needs and expectations and to translate them into a legal provision that is 'adequate' for both sides.

3. Building blocks for a model of adequate third-party liability for TMOs: 'going from trust to Warranted Trust'

Given the undesirable developments of TMO practice, the creation of an adequate tortious liability system for TMOs is recommendable for the good of the e-society. In the present situation, the law can provide a form of hedging against the risk that the trust of e-consumers will be misplaced, especially liability law, which offers remedies for breaches of trust.[757] There are several advantages to introducing an adequate tortious liability system for TMOs. Only to mention some of them, liability law can lay down minimum requirements for the protection of e-consumers that cannot be derogated from contractual clauses. As Chris Witting aptly described in his book *Liability for Negligent Misstatements*, in a service-based economy, where individual's decisions are often made with the assistance of experts' advice or information, '[l]iability in negligence ensures that redress is available for the provision of advice or information which is inaccurate or misleading as a result of a failure to take proper care. It permits the commencement

[755] See Chapter 4, Section 2 *supra*. Pacini, C. & Sinason, D. (1999), p. 482; Houston, R.W. & Taylor, G.K. (1999); Epstein, M.J. & Geiger, M.A. (1994), pp. 60-66; Porter, B. (1993), pp. 49-68; Hanks, J. (1992), pp. 33 et seq.; Costello, J. (1991), p. 267; Chaffee, N. (1988), p. 882; Liggio, C. (1975), p. 42.

[756] See Chapter 3, Subsection 5.5 *supra*.

[757] See Schellekens, M. & Prins, C. (2006), p. 49.

of actions not only by parties in direct contractual relations with the statement-maker, but also by those who are in neither direct nor contractual relations with the statement-maker.'[758] Furthermore, the legalization of a relationship can help in transforming trust into a 'commodity [that serves as] an entity that is familiar and subject to pressure to conform to established standards [as a] dependable anchor for easier and more trusting relations as trust becomes routinely and predictably available, formalised, and standardised.'[759] Moreover, TMOs will possibly consider the legal sanctions and make betrayal of trust less likely to happen *ex ante*. In fact, liability rules could be used as a springboard for the creation of sound trust in the relationship between e-consumers and TMOs.

As none of the laws applicable to TMOs sets forth specific rules on TMO liability, the system will have to be created almost from scratch.

So arguments to answer the third and last fundamental question of this book: *What will an adequate TMO third-party liability system be?* will be unfolded through Sections 3, 4, and 5.

3.1 Basis of liability

The key aim of this effort is to ensure that e-consumers receive at least the same protection in the online world as in the more traditional paper-based and oral forms of communication.[760] TMO offline equivalents have been identified in surveyors, auditors, and accountants.[761] More precisely, following the principle of functionality, trustmarks will be seen as information on somebody or something to be relied upon by others. Thus, TMOs will be seen as professionals who provide information on their clients or their clients' practice, to be relied upon by third parties (e.g., e-consumers). Surveyors, auditors, and accountants can also be seen as professionals who provide information on somebody or something, to be relied upon by oth-

---

[758] Witting C. (2004), p. 3.

[759] Sitkin, S.B. (1995), p. 185. See also Cross F. (2005), p. 1460. 'law may function to expand the radius of trust, make it more likely that individuals will trust those outside their inner circle of family and close friends.'

[760] This aim is often referred to as 'what applies offline should apply online'. See Schellekens, M. (2006), pp. 51-76; Koops, B.J. et al. (2000).

[761] See Chapter 1, Subsection 3.1 *supra*.

ers. The comparison among TMOs, surveyors, auditors, and accountants is further specified in Table 6.

	TMOs	*Surveyors*	*Auditors/Accountants*
Independent and trustworthy info	Trustmarks	Valuations	Reports
On the quality of goods or practice	Security, privacy and business practices	Real estate	Accounts
To be relied upon by third parties	E-consumers	Purchasers	Shareholders/investors

Table 6. Table of comparison

Thus, the core of TMO liability has to be based on the same principles of liability of surveyors, auditors, and accountants. However, the legal analysis carried out in the previous chapters has shown that the principles of liability of surveyors, auditors, and accountants taken as such do grant e-consumers protection only in theory. In fact, due to the specific features of TMOs services, it will be very unlikely for e-consumers to obtain from TMOs compensation for the damages suffered by relying on inaccurate trustmarks.[762] Hence, for the TMO third-party liability system to work as a deterrent to TMO untrustworthy practice, the principles of liability of surveyors, auditors, and accountants have to be tailored to the specific TMO practice. This will be done hereunder by developing the concept of 'adequacy'.

3.2 The theory of Warranted Trust applied to define the concept of 'adequacy'

The concept of 'adequacy' will be defined by applying the theory of Warranted Trust to the relationship between e-consumers and TMOs – which, as it will be shown, aims to protect trustor reliance on trustee by implementing a regulative framework which takes into consideration the interest of both parties and the influence of the specific context in which the trust

[762] See Chapter 8, Section 2 *supra*.

relationship develops – and considering the social, economic, and political value of trustmarks.

> 'Trust (...) is warranted if [the trustees] have good intentions, are competent and work according to a regulative framework and the regulative framework is adequate' (Van Gorp 2005).[763]

If Van Gorp's definition is applied to the relationship between e-consumers and TMOs, the result will be:

> 'Trust is warranted if TMOs have good intentions, are competent, work according to a regulative framework, and the regulative framework is adequate.'

Four elements can be identified in the definition of Warranted Trust:

1. TMOs' good intention;
2. TMOs' competence;
3. TMOs' compliance with the regulative framework; and
4. Adequate regulative framework for TMOs.

Comparing the definition of Warranted Trust to a widely recognised definition of trust – in trust, we rely on the goodwill of someone else – it results that Warranted Trust elements 1, 2, and 3 are all expressions of TMOs' goodwill: goodwill in general, goodwill in entering the trustmark business with enough competence, and goodwill in complying with the regulative framework. Element 4, 'adequate regulative framework', is the only one independent from TMOs' goodwill, so it can be pointed out as the element which differentiates Warranted Trust from trust. Moreover, besides being the characteristic mark of Warranted Trust, 'adequate regulative framework' is an objective element which can be rationalised and influenced by regulations, as opposed to goodwill and related matters, which are by definition subjective and can therefore neither be rationalised nor directly influenced. Nonetheless, it will be argued below that an adequate liability system for TMOs can indirectly stimulate TMOs' good intention, competence, and compliance with TMO duties and obligation.

---

[763] Van Gorp, A. (2005), p. 41.

When it comes to the regulation of the Internet, the success of a regulative framework depends mainly on the possibility of enforcing it.[764] Liability rules are the element of the regulative framework which guarantees its enforcement – at least in theory.[765]

Within the regulative framework, the focus will be on the liability system. Thus, for the scope of the present analysis, the definition of Warranted Trust can be rewritten as follows:

> 'Trust is warranted when an adequate liability system is in place.'

Now the question that needs to be answered is: What does 'adequate' mean?

The definition of 'adequate' will be elaborated taking into consideration the trust relationship between e-consumers and TMOs, that is, the concrete situation to which the liability system will apply. More precisely, the concept of adequacy will be defined and refine through a three-stage analysis of the trust relationship between e-consumers and TMOs.

3.2.1 *Baier's trust stage*

While people have a generalised understanding of the concept of trust, it is not readily amenable to clear definition. Research has at least implicitly accepted a definition of trust as a belief, attitude, or explanation concerning the likelihood that the actions of another individual, group or organisation will be acceptable or will serve the actor's interests. For the purpose of this analysis, the definition of the philosopher Annette Baier will be used: a three-place predicate (A trusts B with valued thing C). 'Valued thing C'

---

764 See, e.g., Prins, C. (2006), pp. 185-187. Illegal filesharing is quite a relevant example of this. Music, films and software's are generally protected by the copyright law in Europe. Filesharing, as a way to circumvent copyright law is bended in all Europe. However, only very rarely have P2P network providers or internet users who illegally shared or downloaded, e.g., copyrighted music, been held liable. Copyright law exists, yet it seems quite difficult to enforce it on the Internet. As a result, illegal filesharing activities proliferate, proving the present regulative framework unsuccessful.

765 In practice, as it was shown in the comparative analysis (Chapters 4, 5, 6, and 7) and in the related conclusions (Chapter 8) that the effectiveness of a liability system depends not only on the existence of a rule but also on the actual possibility that the aggrieved party has to activate it (e.g., to prove the necessary elements to make the third party liable) and on favourable policy arguments.

here means 'things the trusting person values'.[766] Consequently, Baier's definition of trust can be rewritten as follows:

> 'A trusts B with things A values. Applying the new definition to the relation between e-consumers and TMOs it becomes: e-consumers trust TMOs with things e-consumers value.'

What are the things e-consumers value? Nowadays, security, privacy, and more generally, fair business practice are the things that e-consumers value.[767] Having defined what e-consumers value is, Baier's definition of trust can be changed accordingly as follows:

> 'e-consumers trust TMOs with security, privacy, and business practice.'

In fact, when e-consumers trust TMOs, by relying on security, privacy, or business practice trustmarks, they expect TMOs to have checked whether e-merchants have adequate security measures, an adequate privacy policy, and a fair business policy in place. Thus, e-consumers expect TMOs to take care of what they value.

As the requirements for an adequate liability system need to be in correspondence with the notion of trust, an adequate liability system can be preliminarily defined: a system which protects what e-consumers value (e.g., security, privacy, and fair business practice).

3.2.2 *Noteboom's trust stage*

For Bart Nooteboom, trust is a four-place predicate:

- someone
- trusts someone (or something)
- in some respect
- depending on conditions such as context of action.[768]

---

[766] Baier, A. (1986), p. 237.
[767] See Chapter 2, Section 2 *supra*.
[768] Noteboom, B. (2002), p. 36.

Nooteboom adds to Baier's definition the 'context of action' element and explicitly mentions that trust can be placed not only in persons but also in organisations. In fact, the first two elements of Nooteboom's definition, 'someone trusts someone', are the same as the first two elements of Baier's definition, 'A trusts B'. Nooteboom only specifies further that trust can be in organisations (e.g., TMOs): 'Of course an organisation itself does not have an intention, but it has interests and can try to regulate the intentions of its workers to serve those interests.'[769] The third element of Nooteboom's definition, 'in some respect', overlaps with Baier's element 'with valued thing C'. So the new element is the fourth one 'the context of action'.

3.2.3 *Bai-boom trust stage*

Before further analysing the 'context of action', Baier's and Nooteboom's definitions of trust will be merged creating the 'Bai-boom definition' and applied to the relations between e-consumers and TMOs as follows:

> 'e-consumers trust TMOs with things e-consumers value (e.g., security, privacy, and fair business practice) depending on conditions such as TMOs' context of action.'

Coming back to the 'context of action', it may play a role in adjusting e-consumers' expectations towards TMOs. More generally, you trust somebody or an organisation to do things to the degree to which he/she/it can influence the situation and has the power to change certain situations for the better.

The next question to be answered is:

What is the TMOs' context of action?

The Internet is the TMOs' context of action, which brings into play specific features, e.g., de-materialisation, internationalisation or de-territorialisation, and technological turbulence, that are to be taken into consideration.[770] De-

---

[769] Id. p. 38.

[770] The three features of the Internet have been already defined in Chapter 2, Section 3 *supra*. However, the explanation of the concepts will be re-proposed here to favour an immediate understanding of them by the reader.

materialisation means that the information is no longer supplied and stored in written form, but electronically (as a series of ones and zeros) instead.[771] Internationalisation or de-territorialisation is the change brought about by the Internet as a means of communication not restricted by geographic borders. Furthermore, apart from the observation that to talk about national boundaries no longer makes sense, it would seem to be very difficult, due to the mobility of the traffic on the Internet and the possibility for the senders to hide their identity, to determine where exactly a particular person or organisation is located or whether an activity is being carried out.[772] Technological turbulence refers to the constant and rapid development of technology.[773] De-materialisation, internationalisation, and technological turbulence are three very much interrelated features of the Internet that greatly challenge the positive outcome of the service undertaken by TMOs, i.e., to check whether e-merchants have adequate security measures and an adequate privacy policy in place and that they do business in a fair way – in other words, to take care of what e-consumers value most. More precisely, the de-materialisation of information, such as e-consumers personal data makes it very difficult to check how e-merchants actually process those data. For example, e-consumer data can be transferred to any place in the world in a matter of one click without leaving any trace of it (see also the internationalisation factor playing a role here). De-materialisation of persons, as the fact that persons such as e-merchants can have multiple virtual identities, makes quite difficult the proper identification of e-merchants – especially small and not really known ones – by TMOs. Should e-merchants not comply with TMO standards, it may become quite difficult to find e-merchants who provided fake contacts to carry out enforcement procedures as they could be located anywhere (see again the internationalisation factor also playing a role). The constant and rapid development of technologies also comes with always new and more sophisticated techniques to breach the security of e-merchant websites. This increases the need for TMOs to frequently change security standards in order to keep them always up-to-date and thus to also check frequently on whether e-merchants comply with them.

---

[771] See Chapter 2, Section 3 *supra*.

[772] Id.

[773] Id.

With the 'context of action' identified and the related impact on TMO practice clarified, the definition of the trust relation between e-consumers and TMOs can be rewritten as follows: e-consumers trust TMOs with things e-consumers value (e.g., security, privacy, and fair business practice) depending on circumstances such as TMOs' context of action (i.e., the Internet).

As a result, an adequate regulative framework can now be defined such as:

> 'a regulative framework which protects what e-consumers value (e.g., security, privacy, and fair business practice) considering the TMOs' context of action (i.e., the Internet).'

Along the same line, 'Warranted Trust' can be rewritten as follows:

> 'Trust is warranted when a liability system which protects what e-consumers value (e.g., security, privacy, and fair business practice) considering the TMOs' context of action (i.e., the Internet) is in place.'

3.3 Social, economic, and political value of trustmarks

In order to build an adequate liability system, the social, economic, and political impact of the rules has to be considered.

As already mentioned,[774] trustmarks are a very valuable means to improve e-society and e-economy, and to simplify e-policy. In fact, not only e-consumers, but also e-merchants and governments can benefit a great deal from trustmarks.

E-consumers are not able to scrutinise the policies of organisations, companies, and other participants of the virtual world without borders.[775] Ideally, through trustmarks, e-consumers can receive a sort of guarantee from an independent third party on the quality of, for example, an e-merchant's business practice, its privacy statement, or the security level of its website. Moreover, trustmarks are very easy to recognise and can improve the perception of e-consumers with regard to potential online business partners, provide 'always-available', independent, and trustworthy information on

[774] See Chapter 2, Subsection 5.1 *supra*.
[775] See Chapter 2, Section 3 *supra*.

e-merchants, and thus enhance e-consumer confidence in online transactions.

If the benefits of trustmarks for e-consumers can be summarised by the concept of a better buying experience, the consequential benefit for e-merchants will be a better selling experience. Given the heavy pressure to which merchants are exposed in the online world, which has been aptly described by Assafa Endeshaw in his paper *The Legal Significance of Trustmarks* as follows: '[w]hile it has taken many international companies a succession of decades to establish trust and confidence in the line of business or type of wares and services they have offered to their customers, the current wave of business start ups on the Internet feel hard pressed to succeed fast in the 'Wild Web' lest they perishes with the same speed that they sprouted',[776] trustmarks can help e-merchants to succeed fast. Through the exhibition of the trustmark on their websites, e-merchants can make some information easily available to e-consumers, increase the chance to win their trust, and eventually do some business with them. Moreover, trustmarks offer e-merchants the chance to self-regulate sectors of their activities, set out their own standards, and thus prevent governments from interfering.

Last but not least, enhancing consumer trust in e-commerce tops the European Union agenda.[777] The reason is very simple: nowadays, trust is money,[778] and with a minority of European citizens engaged in e-commerce activities,[779] there is still a long way to go towards reaping the full benefits of e-commerce. Lack of trust is the first reason why consumers do not buy online, given the presence of Internet access and interest in purchasing anything online.[780] Trustmarks are a means of self-regulation which aims to

[776] Endeshaw, A. (2001), p. 204.

[777] See, e.g., the recent study for the European Commission Increasing Trust and Confidence of Consumers in the Information Society (forthcoming); *Eurobarometer* (2004); i2010 Five-year strategy to boost digital economy <http://europa.eu.int/information_society/eeurope/i2010/index_en.htm>; the set-up of a dedicated European Agency on information security (ENISA) also highlights the political significance of information security and the need to strive for greater cooperation across EU Member States as well as internationally, see <http://www.enisa.eu.int/>; see also the very recent document produced by the ENISA *Information Security Awareness Initiatives: Current Practice and Measurement of Success* (July 2007).

[778] See, e.g., Valente, K. (2001).

[779] *Eurobarometer* (2004), p. 3.

[780] *Eurobarometer* (2004), pp. 9-11.

enhance e-consumers' trust in online communications. Therefore, trustmarks can give governments the chance to stay out of some sectors of Internet regulation, thus relieving them of an additional financial burden. Concurrently, trustmarks can help boost e-commerce and bring governments the related revenues.

3.4 A model for an adequate TMO liability system

An adequate liability system for TMOs should be based on the liability rules that apply to surveyors, accountants, and auditors which then will have to be tailored to TMOs in order to:

1. effectively protect what e-consumers' value and the related expectations that e-consumers put into their trust relationship with TMOs;
2. take into account the difficulties that TMOs face by operating in a context of action such as the Internet;[781] and
3. bring TMO practice up to the quality level which will give trustmarks the opportunity to extend their potential benefits to social, economic, and political levels.[782]

The model for an adequate liability system is summarised in Table 7.

Legal basis	Surveyors', auditors' and accountants' liability
Adequacy	What e-consumers value
	TMOs' context of action
	Social, economic, and political value of trustmarks

Table 7. Model for an adequate TMO liability system

4. COMPILING THE MODEL: A CONCRETE PROPOSAL FOR AN ADEQUATE TMO THIRD-PARTY LIABILITY SYSTEM

The liability model will be compiled in three steps: first, third-party liability will be imposed on TMOs in order to protect what e-consumers value;

[781] See *supra* Subsection 3.2.3 *Bai-boom trust stage.*
[782] See Chapter 2, Section 5 *supra.*

second, TMO third-party liability will be mitigated keeping into consideration the difficulties faced by TMOs related to the context of action in which they operate; and third, the possible impact of such TMO third-party liability system will be assessed against the social, economic, and political value of trustmarks.

Before starting, it is helpful to recall the typical TMO third-party liability case.

An e-consumer relying on a security, privacy, or business practice trustmark on an e-merchant website decides to interact with such e-merchant. The e-consumer purchases, for example, goods from the e-merchant website or subscribes for an e-merchant online service. In order to complete the transaction, the e-consumer provides the personal data the e-merchant asked for (e.g., name, surname, data of birth, phone number, address, e-mail address, his interests, his purchase preferences) together with the relevant payment details (e.g., credit card number, bank details).

In fact:

a) the e-consumer never receives the good or the service he paid for;
b) without the e-consumer's prior consent, his personal data are processed by the e-merchant for purposes other than the fulfilment of the relevant contractual obligations (e.g., used for profiling- and marketing-related purposes, shared or sold to third parties) and eventually, the e-consumer starts to receive unsolicited marketing e-mails and phone calls from the e-merchant and/or by third parties;
c) the e-consumer's payment details are used directly by the e-merchant to defraud the e-consumers; they are shared or sold by the e-merchant to third parties who ultimately frauds the e-consumer; or stolen during the transaction or from the e-merchant's client database by a cunning third party who takes advantage of the poor security of e-merchant's IT infrastructure.

In any or a combination or all of the above cases, the e-consumer will suffer damage. The damage can range from the violation of the e-consumer's privacy and data protection right[783] to pure economic loss.

[783] Generally recognised in all Europe through the implementation of the Directive 95/46/EC of the European Parliament and of the Council of 24 October 1995 on the protection of individuals with regard to the processing of personal data and on the free movement of such data, *OJ* L 281, 23.11.1995, p. 31.

4.1 Step 1: What e-consumers value

It was already pointed out that, in order to substantiate a claim for damage directly against TMOs, e-consumers generally will have to prove that: (a) the damage occurred to them; (b) TMO fault in the issuance of the trustmark; and (c) the causal link between TMO fault and the damage occurred. In almost all the legal systems analysed, in order to meet the causation requirement, the e-consumer would have to prove his reasonable reliance on the trustmark and that it was foreseeable for the TMO that its carelessness might cause damage to the e-consumer.[784] To provide all these pieces of evidences is already very difficult because of e-consumers' lack of information or TMO practice and, more precisely, fair security, privacy, and business standards.[785] However, even if e-consumers manage to prove all this, they could still see their claim rejected by means of adverse policy arguments leading to a *de facto* absence of liability.[786] The effort is thus here directed to create a system which, in the first place, effectively protects what e-consumers value and the related expectation they put into their trust relationship with TMOs.

As already shown, e-consumers value high e-merchant website security, and fair privacy and business practice. In fact, they trust TMOs. Moreover, e-consumers trust TMOs by relying on security, privacy or business practice trustmarks; they expect TMOs to have checked whether e-merchants have in place adequate security measures and privacy policies, and conduct fair business activities. This way, e-consumers expect TMOs to take care of what they value. From a strictly legal point of view, in order to protect what e-consumers value, the TMOs that represent themselves as reliable sources of information on e-merchant security, privacy, and business practice to e-consumers, should accordingly be held liable if the information they provided through trustmarks is not accurate.

E-consumers shall first prove the damage that occurred to them. Second, they shall prove their reliance on the relevant trustmarks, whereas the reasonableness of their reliance and the foreseeability from the TMOs' side that their carelessness may cause damage to e-consumers will be assumed.

---

[784] See Chapter 8, Section 2 *supra*.
[785] See Chapter 2, Section 4 *supra*.
[786] See Section 3 *supra*.

The reason for this assumption rests in the inner scope of TMO activities. The very task of TMOs is to enhance the level of trust in electronic communications and transactions by issuing trustmarks, which are made to be relied upon. This is the main function of TMOs, the reason why they were created.[787] Moreover, the reason why e-consumers rely on trustmarks is that they are issued by TMOs, which are also defined as TTPs – parties that can be trusted and relied upon.[788] There seems to be enough reasons to assume that e-consumers' reliance on trustmarks is always reasonable. Moreover, if reliance is always reasonable, it goes without saying that TMOs should foresee that an inaccurate trustmark may cause damage to e-consumers by TMOs. However, in order to guarantee an effective protection of what e-consumers value, given a loss suffered by e-consumers because of their reliance on inaccurate trustmarks, the traditional principle of *actori incumbit probatio,* according to which it is the plaintiff who shall prove the defendant's negligence, should be reversed. In fact, because of the general lack of competence that e-consumers have on both TMO practice and, more precisely, fair security, privacy, and business standards[789] it will be very difficult – if not impossible – for them to prove TMO negligence. Therefore, in order to avoid the problem for e-consumers of proving the failure of the TMOs to take reasonable care in providing their services, there should be created a presumption of TMO negligence. Nevertheless, e-consumers will still have to prove that the presence of the trustmark on the e-merchant's website determined their decision to transact with the latter.

4.2 Step 2: TMO context of action

However, considering the difficulties faced by TMOs by providing their services on the Internet due to the de-materialisation of information and identities, internationalisation, and technological turbulence,[790] such presumption of liability should not be an absolute one, which would lead to a regime of strict liability for TMOs. Instead, it should be possible for TMOs to rebut the presumption of liability by proving that they actually act with

---

[787] See Chapter 2, Section 5 *supra.*
[788] See Chapter 1, Section 1 *supra.*
[789] See Chapter 2, Section 4 *supra.*
[790] See *supra* Subsection 3.2.3 *Bai-boom trust stage.*

reasonable care. In other words, it is only the burden of proof concerning the fact that TMOs provided their services with reasonable care that should be shifted onto the TMOs that have the necessary know-how to investigate the matter.

Normally, to act with reasonable care means to perform the service with a diligence and competence appropriate to the profession. However, as the TMO profession is still at the early stages in Europe, standards of reasonable care have not been established. While it is possible to find the definition of 'reasonable care' in the code of conduct of other more established professions (e.g., surveyors, accountants, and auditors), a code of conduct of TMOs does not exist. TMO standards of care should be set by the European legislator – given the 'internationalisation' of online services it should be even better to have an international regulatory body setting worldwide standards for TMOs. Being official standards, whether developed by a European or international recognised body, they will be 'open' and 'public', which means that participation in their development is open to all the stakeholders and the results of the standardisation are publicly available.[791] These standards could actually be based on the key elements of a trustworthy certification practice which have been pointed out at the beginning of this book,[792] which are: (1) certifier independency; (2) impartiality in the auditing procedure; (3) active monitoring of the certified company; (4) certifier enforcement power; and 5) certifier accountability. However, until the recommended issuance of official TMO standards, standards could also be established by analogy with professionals comparable to TMOs, i.e., surveyors, auditors, and accountants, and by checking whether the TMOs fulfilled the duties and obligations related to the service they provided. These duties and obligation are set forth in the contractual clauses or notices which are displayed somewhere on the TMO's website,[793] so in a sense, the contract does define the tort'.[794]

Moreover, considering again the difficulties faced by TMOs by providing their services on the Internet, TMOs shall be given the possibility of

---

[791] For a definition of official standards (as opposed to *de facto* standards) see *supra* Chapter 3, Subsection 2.1.2 *Official and* de facto *standards*.

[792] See Chapter 3, Section 3 *supra*.

[793] E.g., in 'Terms and Conditions', in the 'Certification Practice Statement', in the Seal Licence Agreement, or in the 'Relying Party Agreement'.

[794] Reed, C. (2000), p. 140.

limiting the scope of the trustmark[795] (e.g., trustmarks which do not guarantee e-merchants' security as a whole, but only that e-merchants have in place some specific security measure regarding payments, or trustmarks which guarantee that e-merchants comply with the relevant rules as far as the transferring of personal data towards third countries is concerned) and their validity period.

In summary, by operating this presumption of TMO liability, the e-consumers shall in fact only prove that damage occurred because they relied on an inaccurate trustmark. The TMOs will then be liable for the accuracy of the information on the trustmark for the whole validity of it unless they prove either that the trustmarks have been used (i.e., relied upon) for purposes that exceed the limits or that they acted with reasonable care in issuing the trustmark and checking e-merchant compliance.

4.3 Step 3: Social, economic, and political value of trustmarks

Official standards for TMO practice, coupled with a liability system which effectively takes into account both e-consumer and TMO interests, will most likely improve the quality of TMO services. On the one hand, official standards will set the basic principles for a trustworthy quality certification practice. On the other hand, such liability system will encourage TMOs' good practice by making them liable if e-consumers relying on inaccurate trustmarks suffer loss and TMOs have not complied, though without overburdening them.

E-consumers, e-merchants, governments and also TMOs will all benefit from a rehabilitation of the trustmark system. E-consumers will have a reliable source of information in order to fill the information gap (also defined as information asymmetry) – already documented in Chapter 2, Section 4 – they have with respect to e-merchant security, privacy, and business practice, and thus overcome this specific barrier to the easier adoption of e-commerce. E-merchants will benefit from renewed confidence among e-consumers in online transactions and, more precisely, through the exhibi-

---

[795] Such limitation has been already granted to CSPs (that are professionals which for the services they provide are comparable by analogy to TMOs – as already showed in Chapter 1, Subsection 3.1) in Article 6 (3) of the Directive of the European Parliament and of the Council of 13 December 1999 on a Community Framework for Electronic Signatures (99/93/EC), *OJ* L 13, 19 January 2000, p. 12.

tion of trustmarks on their website, e-merchants can make some information easily available to e-consumers, increase the chance of winning their trust, and eventually doing some business with them. On the side of governments, trustmarks are a means of self-regulation,[796] which aims to enhance e-consumer trust in online communications and transactions. Hence, trustmarks can give governments the chance to stay out of the related sectors of Internet regulation, relieving them of an additional financial burden. Concurrently, trustmarks can help boost e-commerce and bring governments the related revenues. Last but not least, given the benefits for e-consumers, e-merchants, and governments, the demand for trustmarks will increase and TMOs will have the chance to expand their business and reap the relative benefits. Hence, such liability system may also work as an incentive which can indirectly stimulate TMO good intentions (i.e., a fair certification practice), enhance TMO competence (i.e., quality standards), and compliance with the certification standards.[797]

Considering the abovementioned social, economic, and political positive impact of rehabilitating TMO practice, the policy arguments which, as already previously pointed out, weigh so much in deciding the outcome of a third-party liability claim[798] should vary accordingly. This means that once the necessary elements for e-consumers to have a cause of action against TMOs are met (i.e., e-consumer damage, factual reliance on trustmarks, and the causation link between the inaccurate trustmarks and the damage e-consumers suffered), the courts should not bar it. Practically speaking, the most recurrent policy arguments concerning unlimited liability (i.e., arguments to limit third-party professional liability for pure economic loss towards an unlimited number of people for an indeterminate amount) may be counterweighted by: (a) maintaining that the damages in a typical third-party liability case may never consist of a large amount; and (b) possibly imposing a compulsory TMO third-party liability insurance. It has been pointed out that such duty was already imposed on CSPs under French law.[799] This way, the relative costs could be passed on to the e-merchants by slightly increasing the price of TMO services. This should not result in

796 See Chapter 2, Section 5 *supra*.
797 See Subsection 3.2 *supra*.
798 See Section 3 *supra*.
799 See Chapter 7, Section 6 *supra*.

an insuperable disincentive because the current fees for trustmarks are quite low.[800]

5. A possible TMO third-party liability provision

Based on the considerations made above and on the experience of CSP liability as set forth in the Electronic Signatures Directive and its implementation in the analysed European legal systems, the proposed model of adequate third-party liability can actually be summarised into the following liability provision:

> *By issuing a trustmark, a Trustmark Organisation is liable for damage caused to any person who relies on such trustmark as to the accuracy of the information represented in the trustmark for the time of its validity, unless the Trustmark Organisation proves that it has not acted negligently.*
> *A Trustmark Organisation may indicate limitations on the scope of a trustmark provided that they are clearly recognisable by third parties. Accordingly, the Trustmark Organisation shall not be liable for damage arising from the reliance on the trustmark for purposes that exceed the limitations related to it.*

6. Possible impacts of the proposed TMO liability provision

It is now interesting to see how the proposed TMO liability provision would work in practice. In the typical TMO third-party liability case, the aggrieved e-consumer will have to prove that he relied on the security, privacy, or business practice trustmark on the website of the e-merchant in order to transact with the latter. Furthermore, the e-consumer will have to prove that

[800] As a matter of completeness, it is worth stressing that introducing compulsory TMO third-party liability insurance may however result in a disincentive for TMOs to perform their services more carefully. In fact, the insurance lowers the TMO liability risk. Moreover, it will also have to be checked whether insurance companies are actually ready to offer unlimited insurance coverage to TMOs and the related costs.

he suffered damage which is related to that transaction. At this point, according to the proposed provision, the TMO will be presumed liable for that damage. In order to rebut such presumption, the TMO will have to prove alternatively that:

a) it checked and monitored the e-merchant's security, privacy, or business practice with due care and issued the relevant trustmark accordingly;
b) by the time the e-consumer relied on the trustmark, it was not valid because it had already expired; or
c) the damage suffered by the e-consumer was not caused by the e-merchant's failure to comply with what is stated in the trustmark (e.g., the trustmark certified e-merchant's security and the damage was caused not because the e-consumer's payment details were stolen during the transaction by a cunning third party who took advantage of the poor security of the e-merchant's IT infrastructure, but because the e-merchant directly used the e-consumer's payment details in order to defraud him).

As already pointed out, the TMO has all the necessary technical knowledge to provide evidence that it did not act negligently in the issuance and monitoring of the trustmark and that the damage was not caused by e-merchant failure to comply with what was stated in the relevant trustmark. Therefore, the TMO would have all the necessary means to rebut the presumption of negligence and prove that it is not liable for the damage suffered by the e-consumer.

It could be argued that such liability system will exponentially increase e-consumers' actions for damage against TMOs. In fact, for any damage e-consumers suffer from a transaction with an e-merchant that has a trustmark, it will be quite easy for e-consumers to trigger the presumption of TMO negligence. In such case, no matter if the TMO may easily prove to have acted with due care, the (legal) costs that the TMO will have to bear in order to face the possible significant number of e-consumers' claims will dramatically increase to the detriment of its business. Although this is theoretically possible, it seems not very foreseeable in practice. Generally, one e-consumer cannot suffer a big loss from transacting with e-merchants. Therefore, if the e-consumer is not quite sure that he has some chances of holding the TMO liable, he will not start a legal action against the latter because of the related costs. It is more realistic to imagine a situation in which more e-consumers would have suffered damage from transacting with the same e-merchant and a class action is proposed. As class actions have only recently entered into European legal systems, it is very difficult to predict their possible impact on TMOs. On the one hand, the damages that could be claimed may reach amounts which can seriously jeopardise TMOs' business. On the other hand, in countries where class actions are already a consolidated practice, it does not happen everyday to see one of them. Thus, the already-mentioned argument, that a class action will not be proposed if there are quite some chances of holding the TMO liable, may be valid. Therefore, a provisional conclusion on this delicate issue could be that the threat of class actions may work as an incentive to develop TMOs' best practice. The proposed liability system may thus work as a deterrent to discourage TMOs from providing their services carelessly without in fact exposing them to an excessive liability risk.

Moreover, it could be argued that the proposed liability provision will fit without too much trouble into the analysed European legal systems. The presumption of TMO negligence together with the limitation of TMO liability to the scope for which the trustmark was issued are in fact borrowed by the CSP liability system which has been implemented in England, Germany, and France.[801] Furthermore, the fact that e-consumers' reliance on trustmarks is deemed always reasonable could, at first sight, create some compatibility problem with the English system. The recipient of information who wants to claim damage suffered from negligent misstatements has to prove his reasonable reliance on the inaccurate statement to fulfil the proximity test.[802] However, it has already been argued that the reliance of the recipient of information is usually considered reasonable when the other proximity requirements are fulfilled.[803] Accordingly, e-consumers reliance will then be reasonable as in the typical TMO third-party liability case where the proximity factors are all satisfied.[804]

---

[801] See Chapter 5, Section 7; Chapter 6, Section 6; Chapter 7, Section 6 *supra*.
[802] See Chapter 8, Section 2 *supra*. See also, e.g., Witting, C. (2004), pp. 321-322.
[803] See Chapter 8, Section 2 *supra*. See also, e.g., Witting, C. (2004), p. 197.
[804] See Chapter 5, Subsection 6.1 *supra*.

7. CONCLUSIVE REMARKS AND RECOMMENDATIONS

Are TMOs liable towards e-consumers who detrimentally rely on inaccurate trustmarks and suffer loss?

In theory, there is enough legal ground to enforce TMOs' third party liability; in practice, however, the chances that TMOs will not be liable towards aggrieved third-party e-consumers who relied on the misleading or false information provided in the trustmark are way greater than chances that TMOs will be held liable. Thus, at present, it is possible to conclude for a *de facto* absence of TMOs liability towards e-consumers in Europe.[805]

Is the present TMO third-party liability system in Europe adequate?

The present TMO third-party liability system is manifestly inadequate because, to use an euphemism, it does not work as a deterrent to discourage TMO untrustworthy practice. In fact, the total lack of specific rules on TMO liability, coupled with the practical impossibility for e-consumers to hold TMO liable in tort or in contract, as shown in this book, creates a sort legal 'immunity' for TMOs which has surely not helped fostering best practice.[806]

What will an adequate TMO third-party liability system be?

An adequate TMO third-party liability sytem would be a system which:

a) effectively protects what e-consumers value and the related expectations that e-consumers put into their trust relationship with TMOs;
b) takes into account the difficulties that TMOs face by operating in a context of action such as the Internet;[807]
c) brings TMO practice up to the quality level which will give trustmarks the opportunity to extend their potential benefits to social, economic, and political levels;[808]

---

805 See Section 3 *supra*.
806 See Section 4 *supra*.
807 See Subsection 3.1, 3.2 and 3.3 *supra*.
808 See Chapter 2, Section 5 *supra*.

and that can be summarised from a strictly-legal perspective in the following liability provision:

> *By issuing a trustmark, a Trustmark Organisation is liable for damage caused to any person who relies on such trustmark as to the accuracy of the information represented in the trustmark for the time of its validity; unless the Trustmark Organisation proves that it had not acted negligently. A Trustmark Organisation may indicate limitations on the scope of a trustmark provided that they are clearly recognisable by third parties. Accordingly, the Trustmark Organisation shall not be liable for damage arising from the reliance on the trustmark for purposes that exceed the limitations related to it.*[809]

Recently, both in Europe and in other countries, there has been quite a lot of talk around the potential impact of trustmarks on e-commerce as means to foster e-merchant best practice and contextually enhance e-consumer trust.[810] This book aims to contribute to the discussion by adding a piece of the puzzle that until now is missing: an analysis of TMO liability. In fact, no study has systematically addressed in depth such issue so far. On the basis of the present contribution, the discussion on TMO liability can be promoted with all the stakeholders in Europe as well as on a global level. In fact, the goal should be to create a global system of TMO liability which can apply to TMOs offering their services all around the world. The European Commission and organisations such as the United Nations Commission on International Trade Law (UNCITRAL)[811] and the Organisation for

---

[809] See Sections 5, 6, and 7 *supra*.

[810] See, e.g., Expert Workshop held at the European Commission on 8 March 2007 on Measure to Increase Trust and Confidence for Consumers in the Information Society at the Expert Workshop <http://ec.europa.eu/information_society/eeurope/i2010/studies/index_en.htm>; <http://ec.europa.eu/information_society/eeurope/i2010/docs/studies/1_cons_trust_ws_cases.ppt>;<http://ec.europa.eu/information_society/eeurope/i2010/docs/studies/2_cons_trust_ws_survey.ppt>; <http://ec.europa.eu/information_society/eeurope/i2010/docs/studies/3_cons_trust_ws_country_reports.ppt>; De Bruin, R. et al. (2005) Final Report. See also See Ko, Y. (2007), pp. 20-22; Trustmark Conference, May 15 and 16, 2007, Mexico City, Mexico; APEC ECSG Data Privacy seminar June 22 and 23, 2007, Cairns, Australia.

[811] <www.uncitral.org/>.

Economic Co-operation and Development (OECD)[812] may actually play a crucial role in taking relevant steps to promote and coordinate the creation of European TMO third-party liability standards and stimulating a global harmonisation on the matter, coordinating their efforts with relevant trustmark initiatives all around the world.[813]

TMO third-party liability calls for a global, active approach so that liability can fulfil a useful role in enforcement of self-regulatory schemes without 'overdoing it'. Neither denying the existence of third-party TMO liability nor exaggerating it will lead to a satisfactory result.

---

812 <www.oecd.org>.

813 E.g., Asia-Pacific Economic Cooperation (APEC) Trustmarks initiatives. See Ko, Y. (2007), pp. 20-22; APEC Electronic Commerce Steering Group: <http://www.apecsec.org.sg/apec/apec_groups/som_special_task_groups/electronic_commerce.html>.

LITERATURE

Akerlof, G.A., 'The Market for "Lemons": Quality Uncertainty and the Market Mechanism', *Quarterly Journal of Economics* 84 (1970)

Allegaert, C. & Tinkelman, D., 'Reconsidering the "Lack of Duty" Defense to State Auditor Negligence Claims', *Journal of Corporation Law* 25 (2000)

Arrow, K.J., 'Uncertainty and the Welfare Economics of Medical Care', *The American Economic Review* 53 (5) (1963)

Astrue, M.J., 'Health Care Reform and the Constitutional Limits on Private Accreditation as an Alternative to Direct Government Regulation', *Law and Contemporary Problems* 57 (4) (1994)

Atiyah, P.S., 'Economic Loss in the United States', *Oxford Journal of Legal Studies* (1985)

Baber, W. et al., 'Client Security Price Reactions to the Laventhol and Horwath Bankruptcy', *Journal of Accounting Research* 33 (1995)

Baier, A., 'Trust and Antitrust', *Ethics* (1986)

Bailey, L.P. (2002), *Miller GAAS Guide 2002: A Comprehensive Restatement of Standards for Auditing, Attestation, Compilation, and Review* (New York: Aspen Law & Business)

Balboni, P., 'Liability of Certification Service Providers towards Relying Parties and the Need for a Clear System to Enhance the Level of Trust in Electronic Communication', *Information & Communications Technology Law* 13 (2004)

Bassenge, P. et al. (2006), *Palandt Bürgerliches Gesetzbuch* (Munich: C. H. Beck)

Bell, J. et al. (1998), *Principles of French Law* (Oxford: Oxford University Press)

Bertrand, A.R. (1992), 'La responsabilité du fait des bases de données en France', in Sieber, U. (ed.) *Liability for On-line Data Bank Services in the European Community*, Vol. 3 of the European Series on Information Law (Cologne, Berlin, Munich: Carl Heyman Verlag KG)

Beyer, O., 'Vertraglicher Drittschuz', *JuS* (1996)

Billiau, M., 'Le tiers à un contrat peut invoquer, sur le fondement de la responsabilité délictuelle, un manquement contractual dès lors qu'il lui a cause un dommage', *La Semaine Juridique Edition Générale* no. 46, 15 November 2006, II (2006)

Brun, M. B., 'Nature et impacts juridiques de la certification dans le commerce électronique sur Internet', *Lex Elettronica* 7 n. 1, été 2001. Centre de recherche en droit public, Université de Montréal. Available at <http://www.lex-electronica.org/articles/v7-1/Brun.htm>

Calegari, M. et al., 'Experimental Evidence of Differential Auditor Pricing and Reporting Strategies', *Accounting Review* 73 (1998)

Canaris, C.W. (1971), *Die Vertrauenshaftung im deutschen Privatrecht* (Munich: Beck)

Canaris, C., 'Norme di protezione, obblighi del traffico, doveri di protezione', *Riv. Cr. Dir. Priv.* 1 (3) (1983)

Cane, P. (1996), *Tort Law and Economic Interests* (Oxford: Carendon Press)

Castell, R. & Pasqualini, F. (1995), *Le commissaire aux comptes* (Paris: Economica)

Caswell, J.A. & Mojduszka, E.M., 'Using Information Labelling to Influence the Market for Quality in Food Products', *American Journal of Agricultural Economics* 78 (1996)

Causey, D., 'Accountants' Liability in An Indeterminate Amount for An Indeterminate Class: An Analysis of Touche Ross & Co. v. Commercial Union Ins. Co.', *Mississippi Law Journal* 57 (1987)

Chaffee, N., 'The Role and Responsibility of Accountants in Today's Society', *Journal of Corporation Law* 13 (3) (1988)

Clarke, R. (7 July 1998), *Senate Legal and Constitutional References Committee. Inquiry into Privacy and the Private Sector*. Available at <http://www.anu.edu.au/people/Roger.Clarke/DV/SLCCPte.html>

Clarke, R. (12 April 1999), *The Legal Context of Privacy-Enhancing and Privacy-Sympathetic Technologies*. Available at <http://www.anu.edu.au/people/Roger.Clarke/DV/Florham.html>

Cooke, J. & Oughton, D. (2000), *The Common Law of Tort* (London: Butterworths)

Compagnie Nationale des Commissaires aux Comptes (1995), *La responsabilité civile du commissaire aux comptes* (Paris: Encuclopédie d'études juridiques)

Costello, J., 'The Auditor's Responsibilities for Fraud Detection and Disclosure: Do the Audit Standards Provide a Safe Harbor?', *Maine Law Review* 43 (1991)

Couret, A. et al. (1995), *La Certification* (Paris: Presses Universitaires de France)

Craig, E. (2005), *The Shorter Routledge Encyclopaedia of Philosophy* (New York: Routledge)

Cross F., 'Law and Trust', *George Town Law Journal* 93 (5) (2005)

De Angelo, L.E., 'Auditor Independence, Low Balling and Disclosure Regulation', *Journal of Accounting and Economics* 3 (1981)

De Bruin, R. et al. (2005), *Analysis and Definition of Common Characteristics of Trustmarks and Web Seals in the European Union*: Final Report + Annexes (2 Vols.)

De Cock Buning, M. et al., 'Consumer@Protection.EU. An Analysis of European Consumer Legislation in the Information Society', *Journal of Consumer Policy* 24 (2001)

De Vrey, R.W. (2006), *Towards a European Unfair Competition Law. A Clash Between Legal Families* (Leiden/Boston: Martinus Nijhoff Publishers)

Deakin, S. et al. (2003), *Markesinis and Deakin's Tort Law* (Oxford: Oxford university Press)

Dean, H. & Biswas, A., 'Third-Party Organization Endorsement of Products: An Advertising Cue Affecting Consumer Pre-purchase Evaluation of Goods and Services', *Journal of Advertising* 30 (2001)

Demogue, R. (1928), *Traité des obligations in general* (Paris) vol. 5

Directive 95/46/EC of the European Parliament and of the Council of 24 October 1995 on the protection of individuals with regard to the processing of personal data and on the free movement of such data, *OJ* L 281, 23.11.1995

Directive 2005/29/EC of the European Parliament and of the Council of 11 May 2005 concerning unfair business-to-consumer commercial practice in the internal market and amending Council Directive 84/450/EEC

Directives 97/7/EC, 98/27/C and 2002/65/EC of the European Parliament and of the Council and Regulation (EC) No 2006/2004 of the European Parliament and of the Council ('Unfair commercial Practices Directive'), *OJ* L 149, 11.6.2005, p. 22

Directive 2000/31/EC of the European Parliament and of the Council of 8 June 2000 on certain legal aspects of information society services, in the Internal Market (Directive on electronic commerce), *OJ* L 178, 17.7.2000, p. 1

Directive 97/7/EC of the European Parliament and of the Council of 20 May 1997 on the protection of consumers in respect of distance contracts, *OJ* L 144 , 4.6.1997, p. 19

Directive of the European Parliament and of the Council of 13 December 1999 on a Community Framework for Electronic Signatures (99/93/EC), *OJ* L 13, 19 January 2000, p. 12

Dobbs, D.B. (2000), *The Law of Torts* (St. Paul MN: West Group)

Driver, A.W. (2006), 'Navigate the Privacy Maze in the U.S. and Abroad', *The Metropolitan Corporate Counsel* July, p. 7. Available at <http://www.metrocorpcounsel.com/current.php?artType=view&artMonth=July&artYear=2006&EntryNo=5301>

Dugdale, A.M. & Jones, M.A. (2006), *Clerk & Lindsdell on Torts* (London: Sweet & Maxwell)

Duthler, A.W. (1998), *Met recht een TTP! Een onderzoek naar juridische modellen voor een Trusted Third Party* (Deventer: Kluwer)

Egger, F. (2002), 'Consumer Trust in E-commerce: From Psychology to Interaction Design', in Prins, J.E.J. et al. (eds.) *Trust in Electronic Commerce. The Role of Trust from a Legal, an Organizational and Technical Point of View* (The Hague: Kluwer Law international)

Endeshaw, A., 'The Legal Significance of Trustmarks', *Information & Communication Technology Law* 10 (2) (2001)

ENISA *Information Security Awareness Initiatives: Current Practice and Measurement of Success* (July 2007). Available at: <http://www.enisa.europa.eu/doc/pdf/deliverables/enisa_measuring_awareness.pdf>

Epstein, M.J. & Geiger, M.A., 'Investor Views of Audit Assurance: Recent Evidence of the Expectation Gap', *Journal of Accountancy* 177 (1994)

Epstein, R.A. (1995), *Simple Rules for a Complex World* (Cambridge, MS., London: Harvard University Press)

Eurobarometer (2004), 'Issues Relating to Business and Consumer e-Commerce', European Commission. The executive summary is available at <europa.eu.int/comm/consumers/topics/btoc_ecomm.pdf>

Eurobarometer (2003), 'Data Protection, European Commission'. Available at <http://www.ec.europa.eu/public_opinion/archives/ebs/ebs_196_data_protection.pdf>.

European Trustmark Requirements (ETR). Available at <http://www.quatro-project.org/files/file/unice-beuc/eConfidence.pdf>

Farrell, H. (2005), 'The Political Economy of the Internet and E-Commerce', in Stubbs, R. & Geoffrey, R.D. (eds.) *Political Economy and the Changing Global Order* (third edition) (Oxford: Oxford University Press). Available at <http://www.henryfarrell.net/internet.pdf>

Feinman, J.M., 'The Significance of Contract Theory', *University of Cincinnati Law Review* 58 (1990)

Feinman, J.M. (1995), *Economic Negligence: Liability of Professionals and Businesses to Third Parties for Economic Loss* (Waltham, MA: Little Brown & Co Law & Business)

Feinman, J.M., 'Relational Contract Theory in Context', *Northwestern University Law Review* 94 (2000)

Feinman, J.M., 'Liability of Accountants for Negligent Auditing: Doctrine, Policy, and Ideology', *Florida State University Law Review* 31 (2003)

Fleming, J., 'Property Damage – Economic Loss: A Comparative View', *LQR* 105 (1989)

Flour, J. & Aubert J.L. (1997), *Les Obligations* (Paris: Colin)

Froomkin, M., 'The Essential Role of Trusted Third Parties in Electronic Commerce', *Oregon Law Review* 75 (1996). Available at <http://www.osaka.law.miami.edu/~froomkin/articles/trusted.htm>

Froomkin, M., 'The Death of Privacy?', *Stanford Law Review* 52 (2000). Available at <http://www.personal.law.miami.edu/~froomkin/articles/privacy-deathof.pdf>.

Garner, B.A. (2004), *Black's Law Dictionary* (St. Paul MN: Thomson West)

Giliker, P. & Beckwith, S. (2004), *Tort* (London: Sweet & Maxwell)

Girot, C. (2000), *User Protection in IT Contracts. A Comparative Study of the Protection of the User against Defective Performance in Information Technology* (Amsterdam: Kluwer Law International)

Goldsmith, R. et al., 'The Impact of Corporate Credibility and Endorser Celebrity on Consumer Reaction to Advertisement and Brands', *Journal of Advertising* 29 (3) (2000)

Gottlieb, G., 'Relationism: Legal Theory for a Relational Society', *University of Chicago Law Review* 50 (1983)
Groupements Français des Fournisseurs d'Information en Ligne (1986), *Les relations contractuelles des producteurs des bases et banques de données – Problèmes juridiques* (Paris: Dalloz)
Guyon, Y. (1994), *Droit des affaires* (Paris: Economica)
Hagen III, W.W., 'Accountants' Common Law Liability to Third Parties', *Columbia Business Law Review* 1 (1988)
Hanks, J., 'The Expectations Gap – the Consumer Angle', *Accountancy* 109 (1992)
Hartherly, D.J. et al. (1992), *The Audit Expectations Gap in the United Kingdom* (London: Research Board of the Institute of Chartered Accountants in England and Wales)
Havighurst, C.C., 'The Place of Private Accreditation among the Instruments of Government', *Law and Contemporary Problems* 57 (4) (1994)
Heydn, T. J. (2005), 'Germany', in Campbell, D. (ed.) *E-Commerce and the law of Digital Signatures* (New York: Oceana Publications, Inc.)
Hill, J. et al., 'Auditing's Emerging Legal Peril Under the National Surety Doctrine: A Program for Research', *Accounting Horizons* 7 (1993)
Hirte, H. (1996), *Berufshaftung* (Munich: C.H. Beck)
Hobbs, J.E. et al., 'Incentive Structures for Food Safety and Quality Assurance: An International Comparison', *Food Control* 13 (2002)
Hodgin, R. (1999), *Professional Liability: Law and Insurance* (London/ Hong Kong: LLP Professional Publishing)
Hodgson, J. (2004), *Tort Law* (Oxford: Oxford University Press)
Houston, R., & Taylor, G., 'Consumer Perceptions of CPA Web Trust assurances: Evidence of an Expectation Gap', *International Journal of Auditing* 3 (2) (1999)
Howells, G. et al. (2006), *European Fair Trading Law. The Unfair Commercial Practices Directive* (Aldershot: Ashgate)
Hoxmeier, J., 'Software Preannouncements and their Impact on Customers' Perceptions and Vendor Reputation', *Journal of Management information Systems* 17 (1) (2000)
Huber, P.W. (1988), *Liability: the Legal Revolution and its Consequences* (New York: Basic Books, Inc.)
Humphrey, C.G., et al., 'The Audit Expectation Gap in Britain: an Empirical Investigation', *Accounting and Business Research* 23 (1993)
Increasing Trust and Confidence of Consumers in the Information Society (forthcoming) Study for the European Commission. For the description of the project see <http://www.ec.europa.eu/information_society/eeurope/i2010/studies/studies_ongoing/index_en.htm>
Information Society (2005), *Scoping Study for the Measurement of Trust in the Online Environment*, DSTI/ICCP/IIS(2005)1/FINAL. Available at <http://www.oecd.org/dataoecd/26/15/35792806.pdf>

Jahn, G. et al., 'The Reliability of Certification: Quality Labels as a Consumer Policy Tool', *Journal of Consumer Policy* 28 (2005)

Jhering, R. (1861), 'Culpa in contrahendo oder Schadenersatz bei nichtigen oder nicht zur Perfektion gelangten Verträgen', in Jhering (ed.) *Jahrbücher für die Dogmatik des heutigen römischen und deutschen Rechts 4*

Kaner, C., 'Liability for Defective Content', *Software QA* 3 (3) (1996). Available at <http://www.kaner.com/pdfs/content.pdf>

Keeton, W.P. et al. (1984), *Prosser and Keeton on Torts* (St. Paul MN: West Group)

Keller, K.L. (1998), *Strategic Brand Management* (Upper Saddle River, NJ: Prentice Hall).

Kennedy, D. (1997), *A Critique of Adjudication (fin de siecle)*, (Cambridge: Harvard University Press)

Khoury, L., 'The Liability of Auditors beyond Their Clients: A Comparative Study', *McGill Law Journal* 46 (2001)

Ko, Y., 'APEC Cross-Border Privacy Rules and Trustmarks: A Step Toward Integrated Electronic Commerce in the Asia-Pacific', *World Data Protection Report* 7 (3) (2007)

Konrad, H. (1978), 'Verkehrspflichten zum Schutz fremden Vermögens', in C. Ficker et al., *Festschrift für Ernst v. Caemmerer zum 70. Geburtstag* (Tübingen: Mohr)

Koops, B.J. et al. (2000), *ICT Law and Internationalisation. A Survey of Government Views. Law and Electronic Commerce* Vol. 10 (The Hague: Kluwer Law International)

Kornblum, J., 'FTC, GeoCities Settle on Privacy', *CNET News.com*, 13 August 1998. Available at <http://www.news.cnet.com/news/0-1005-200-332199.html>

Kovar, S. et al., 'Consumer Responses to the CPA WebTrust Assurance', *Journal of Information Systems* 14 (1) (2000)

Kreuzer, K.F., 'Entscheidungen Bürgerliches Recht', *JZ* (1976), p. 778

Lafferty, B. et al., 'The Dual Credibility Model: The Influence of Corporate and Endorser Credibility on Attitudes and Purchase Intention', *Journal of Marketing Theory and Practice* 20 (3) (2002)

Lala, V. et al., 'The Impact of Relative Information Quality of e-Commerce Assurance Seals on Internet Purchasing Behavior', *International Journal of Accounting Information Systems* 3 (4) (2002)

Larenz, K. (1956), *Schuldrecht I* (Munich: C.H. Beck)

Larenz, K & Canaris, C. W. (1994), *Lehrbuch des Schuldrechts*, (Munich: C.H. Beck)

Le Tourneau, P. & Cadiet, L. (2002), *Droit de la Responsabilité et des Contracts* (Paris: Dalloz)

Lemaignan, A. & Nechelis, D., 'Expert-comptable', *JurisClasseur Responsabilité civile et Assurances* 376 (2005)

Lemos, R., 'Can You Trust TRUSTe?', *ZDNet News*, 2 November 1999. Available at <http://www.news.zdnet.com/2100-9595_22-516377.html>.
Lewis, C.T. (1996), *A Latin Dictionary* (New York: Oxford University Press)
Liggio, C., 'Accountants' Civil Liability: Through the Looking Glass and Other "Impossible Things"', *Litigation* 1 (1975)
Linant de Bellefonds, X. & Hollande, A. (1992), *Contrats informatiques et télématiques* (Paris: Delmas)
'Listening to the Internet. Internet trends: Companies are eavesdropping on online discussion forums to find out what their customers really think about them', *The Economist*, 9 March 2006
Lorenz, W. (1973), 'Das Problem der Haftung fur primare Vermogenschaden bei der Erteilung einer unrichtiger Auskunft', in G. Paulus et al., *Festschrift für Karl Larenz* (Munich: Beck)
Lucas A. (1987), *Le droit de l'informatique*, (Paris: PUF)
Lunney, M. & Oliphant, K. (2007), *Tort Law Text and Materials* (Oxford: Oxford University Press)
Macneil, I.R. (1980), *The New Social Contract: An Inquiry Into Modern Contractual Relations* (New Haven: Yale University Press)
Macneil, I.R., 'Values in Contract: Internal and External', *Northwestern University Law Review* 78 (1983)
Magnus, U. (2001), 'Germany', in Magnus, U. (ed.) *Unification of Tort Law: Damages* (The Hague/London/Boston: Kluver Law International)
Makkawi, B. & Schick, A., 'Are Auditors Sensitive Enough to Fraud?', *Managerial Auditing Journal* 19 (2003)
Malaurie, P. & Aynès, L. (1992), *Cours de droit civil. Les obligations* (Paris: Cujas)
Mann, R.J. & Belzey, S. R., 'The Promises of Internet Intermediary Liability', *Wm. & Mary L. Rev.* 47 (2005)
Markesinis, B. S. & Uberath, H. (2002), *The German Law of Torts: A Comparative Treatise* (Oxford and Portland, Oregon: Hart Publishing)
Mauldin, E. & Arunachalam, V., 'An Experimental Examination of Alternative Forms of Web Assurance for Business-to-Consumer e-Commerce', *Journal of Information Systems* 16 (1) (2001). Available at <http://www.allbusiness.com/sales/internet-e-commerce/839064-1.html>
Mazeau, H. et al. (1965), *Traité théorique et practique de la responsabilité civile délictuelle et contratuelle* (Paris: Éditions Montchrestien)
McCarthy, J., 'Is TRUSTe Trustworthy?', *The Ethical Spectacle*, September 1998. Available at <http://www.spectacle.org/998/mccarthy.html>
McCarthy, J., 'TRUSTe Decides Its Own Fate Today', *Slashdot*, 8 November 1999. Available at <http://www.slashdot.org/yro/99/11/05/1021214.shtml>
McKnight, H. et al., 'Developing and Validating Trust Measures for e-Commerce: An Integrative Typology', *Information System Research* 13 (3) (2002). Available at <http://www.msu.edu/~mcknig26/Measures.pdf>

McWilliams, B., 'Real Hit With Another Privacy Lawsuit', *Internet news.com,* 10 November, 1999. Available at <http://www.internetnews.com/streaming-news/article/0,1087,8161_236261,00.html>

Medicus, D. (1986), 'Entdeckungsgeschichte der culpa in contrahendo', in H. P. Benöhr et al., *Festschrift für Max Kaser zum 80. Geburtstag* (Wien, Köln, Graz)

Menon, K. & Williams, D., 'The Insurance Hypothesis and Market Prices', *The Accounting Review* 66 (1994)

Miller, R. & Young, M., 'Financial Reporting and Risk Management in the 21st Century', *Fordham Law Review* 65 (1997)

Molland, M., 'Privacy Litigation in the United States — The Cookie Class Actions', *Canadian International Lawyer: Revue de Droit International* 4 (2001)

Nicoll, C. et al. (2003), *Digital Anonymity and the Law. Tension and Dimension* (The Hague: TMC Asser Press)

Nordquist, F. et al., *Trusting the Trustmark?*, 17th Bileta Annual Conference, Amsterdam, The Netherlands, 5-6 April 2002. Available at <http://www.bileta.ac.uk/Document%20Library/1/Trusting%20the%20Trustmark.pdf>.

Noteboom, B. (2002), *Trust* (Cheltnham, UK; Northampton, MA, US: Edward Elgar)

Nussbaum, B., 'Can You Trust Anybody Anymore?', *Business Week*, January 28 (2002). Available at <http://www.businessweek.com/magazine/content/02_04/b3767701.htm>

Odom, M. et al., 'Web Assurance Seals: How and Why They Influence Consumers' Decisions', *Journal of Information Systems* 16 (2) (2002). Available at <http://www.goliath.ecnext.com/coms2/gi_0199-2268451/Web-assurance-seals-how-and.html>

Open Letter 1999/11/1 to RealNetworks, RealNetworks' Privacy Intrusion, *Junkbuster* <http://www.junkbusters.com/ht/en/real.html>

Organisation for Economic Co-operation and Development – Working Party on Indicators for the Information Society (2005) Scoping *Study for the Measurement of Trust in the Online Environment*, DSTI/ICCP/IIS(2005)1/FINAL. Available at <http://www.oecd.org/dataoecd/26/15/35792806.pdf

Organisation for Economic Co-operation and Development – Working Party on Indicators for the Information Society (2008) *Measuring Security and Trust in the Online Environment: A View Using Official Data* DSTI/ICCP/IIS(2007)4/FINAL. Available at: <http://www.oecd.org/dataoecd/47/18/40009578.pdf>

Pacini, C. & Sinason, D., 'Auditor Liability for Electronic Commerce Transaction Assurance: the CPA/CA WebTrust', *American Business Law Journal* 36 (1999)

Pacini, C. et al., 'Commonwealth Convergence Toward a Narrower Scope of Auditor Liability to Third Parties for Negligent Misstatements', *Abacus* 38 (3) (2002)

Palmer, J. et al., 'The Role of Intermediaries in the Development of Trust on the WWW: The Use and Prominence of Trusted Third Parties and Privacy Statements', *Journal of Computer Mediated Communication* 5 (3) (2000). Available at <http://www.jcmc.indiana.edu/vol5/issue3/palmer.html>

Panttaja, R., 'Accountants' Duty to Third Parties: A Search for a Fair Doctrine of Liability', *Stetson Law Review* 23 (1994)

Parker, A., 'We have to prove our own quality', *Financial Times*, 21 July 2005

Perlman, H.S., 'Interference with Contract and other Economic Expectancies', *University of Chicago Law Review* 61 (1982)

Pierce, B. & Sweeney, B., 'Cost-Quality Conflict in Audit Firms: An Empirical Investigation', *European Accounting Review* 13 (2004)

Pinna, A. (2003), *The Obligations to Inform and to Advise* (The Hague: Boom Juridische uitgevers)

Ponthier, J., 'La certification outil de la modernité normative' , *Dalloz-Sirey* 30 (1996)

Porter, B., 'An Empirical Study of the Audit Expectation-Performance Gap', *Accounting and Business Research* 24 (1993)

Powell, J.L. & Stewart, R. (2007), *Jackson & Powell on Professional Liability* (London: Sweet & Maxwell)

Principles for e-Commerce Codes of Conduct – Second Draft (The E-Confidence Initiative Working Documents)

Prins, C. (2006), 'Should ICT Regulation Be Undertaken at an International Level?', in Koops, B. J. et al. (eds.) *Starting Point for ICT Regulation* (The Hague: T M C Asser Press)

'Psst, wanna buy a cheap bracelet?', *The Economist*, 1 July 2004. Available at <http://www.economist.com/research/articlesBySubject/displayStory.cfm?story_ID=2877169&subjectid=1198563>

Rae, A. et al. (1995), *Software Evaluation for Certification. Principles, Practice and Legal Liability* (London: McGraw Hill Book Company)

Rafter, M., 'Trust or Bust?', *The Standard* (2000). Available at <http://www.webtrust.fr/Apropos/presse/article%20critique%20sur%20truste.pdf>

Ralf, M. (2006), 'The Functional Method of Comparative Law', in Reimann, M. et al. (eds.) *The Oxford Handbook of Comparative Law* (Oxford: Oxford University Press)

Raspante, J.F. & Rosario, R., 'Understanding and Minimizing CPA Liability', *CPA Journal* (2001)

Reed, C. (2000), *Internet Law: Text and Materials* (London/Edinburgh/Dublin: Butterworths)

Regoli, N.L., 'Indecent Exposures in an Electronic Regime', *Federal Communications Law Journal* 54 (2002). Available at <http://www.law.indiana.edu/fclj/pubs/v54/no2/Regoli.pdf>

Ricci, A., Pelino, E., Balboni, P. and Cunegatti, B. (2008), 'Il diritto al nome e il diritto all'anonimato su Internet', in Finocchiaro, G. (ed.) *Diritto all'anonimato. Anonimato, nome e identità personale* (Padova: Cedam)

Riegelsberger, J. & Sasse, M.A., 'Trust me, I'm a .com. The Problem of Reassuring Shoppers in Electronic Retail Environments', *Intermedia* 28 (4) (2000). Available at <http://www.cs.ucl.ac.uk/staff/A.Sasse/trustme.htm>

Rogers, W.V.H. (2006), *Winfield and Jolowicz on Tort* (Lodon: Sweet & Maxwell)

Rothery, B. (1996), *Standards and Certification in Europe* (Aldershot: Gower Publishing Limited)

Schlechtriem, P., *Civil Liability for Economic Loss Germany*, 15th International Congress of comparative Law Bristol, England, 26 July to 1 August, 1998. Available at <http://www.jura.uni-freiburg.de/einrichtungen/gfr/Bristol/Schlechtriem/schlechtriem.pdf>.

Schlechtriem, P. (1999), 'Schutzpflichten und geschützte Personen', in Beuthien et al. (eds) *Festschrift für Dieter Medicus*, (Cologne: Heymanns)

Schellekens, M. & Prins, C., 'Unreliable Information on the Internet: A Challenging Dilemma for the Law', *Information, Communications and Ethics in Society* 4 (1) (2006)

Schellekens, M. (2006), 'What Holds Off-Line, also Holds On-Line?', in Koops, B.J. et al. (eds.) *Starting Point for ICT Regulation* (The Hague: T.M.C. Asser Press)

Scherl, J.B., 'Evolution of Auditor Liability to Noncontractual Third Parties: Balancing the Equities and Weighing the Consequences', *American University Law Review* 44 (1994)

Siliciano, J.A., 'Negligent Accounting and the Limits of Instrumental Tort Reform', *Michigan Law Review* 86 (1988)

Sitkin, S.B. (1995), 'On the Positive Effect of Legalization on Trust', in R. Lewicki, B. Sheppard, & R. Bies (eds.), *Research on Negotiation in Organisations* Vol. 5 (Greenwood: JAI)

Sonnenschein, J., 'Der Vertrag mit Schutzwirkung fur Dritte – und immer neue Fragen', *JA* (1989)

Spar, D.L. (2001), *Ruling the Waves: Cycles of Discovery, Chaos, and Wealth from the Compass to the Internet* (New York: Harcourt Hardcovers)

Spence, M., 'Informational Aspects of Market Structure: An Introduction', *Quarterly Journal of Economics* 90 (1976)

Sporleder, T.L. & Goldsmith, P.D., 'Alternative Firm Strategies for Signalling Quality in the Food System', *Canadian Journal of Agricultural Economics* 49 (2001)

Starck, B. et al. (1996), *Obligations. 1) La responsabilité délictuelle* (Paris: Litec)

Stice, J.D., 'Using Financial and Market Information to Identify Pre-engagement Factors Associated with Lawsuits Against Auditors', *The Accounting Review* 66 (1991)

Stuurman, K. (1990), 'Legal Aspects of Standardization and Certification of Information Technology and Telecommunications: An Overview', in Kaspersen, H.W.K. & Oskamp, A. (eds.) *Amongst Friends in Computers and Law* (Deventer: Kluwer Law and Taxation Publisher)

Tancelin, M. (1993), *Des obligations: L'acte illégitime et les modes d'exécution* (Montreal: Wilson & Lafleur)

Terré, F. et al. (2002), *Droit Civil: Les Obligations* (Paris: Dalloz)

Thomas, C.W., 'The Rise and Fall of Enron', *Journal of Accountancy* 193 (4) (2002)

Tunc, A., 'La distinction des obligations de résultat et des obligations de diligence', *J.C.P.* (1945)

Valente, K., 'Keeping Promises: Backing up your Promises of Reliability with Proof is Becoming a New Economy Critical Success Factor', *Intelligent Enterprise* (2001). Available at: <http://www.intelligententerprise.com/010613/trust1_1.jhtml;jsessionid=CVK11KPCSJIVEQSNDLPCKHSCJUNN2JVN>

Van Boom et al. (2004), *Pure Economic Loss* (Vienna: SpringerWienNewYork)

Van Dam, C. (2006), *European Tort Law* (Oxford: Oxford University Press)

Van Gerven, W. et al. (2000), *Cases, Materials on National, Supernational and International Tort Law* (Oxford and Portland Oregon: Hart Publishing)

Van Gorp, A. (2005), *Ethical Issues in Engineering Design, Safety and Sustainability* (Eindhoven: Simon Stevens Series in the Philosophy of Technology)

Van Klink, B.M.J. & Prins, J.E.J. (2002), *Law and Regulation: Scenarios for the Information Age*, (Amsterdam / Berlin / Oxford / Tokyo / Washington, DC: IOS Press)

Viney, G., 'La responsabilité des prestataires de conseil', *J.C.P.* (1975)

Viney, G. (1995), *Traité de Droit Civil: Introduction à la Responsabilité* (Paris: Librarie Générale de Droit et de Jurisprudence)

Viney, G. (1998), *Traité de Droit Civil: Les Conditions de la Responsabilité* (Paris: Librarie Générale de Droit et de Jurisprudence)

Viney, G. & Jourdain, P. (2001), *Traité de Droit Civil: Les Effects de la Responsabilité* (Paris: Librarie Générale de Droit et de Jurisprudence)

Viney, G., 'La responabilité du débiteur à l'égard du tiers auquel il a cause un dommage en manquant à son obligation contractuelle', *Recueil Dalloz 2006 Jurisprudence* (2006)

Vinten, G., 'Enronitis – Dispelling the Disease', *Managerial Auditing Journal* 18 (2003)

Von Bar, C. (1981), 'Deliktsrecht', in Bundesministerium der Justiz (ed.) *Gutachten und Vorschläge zur. Überarbeitung des Schuldrechts*, vol. 2 (Cologne: Bundesanzeiger Verlagsges.)

Von Bar, C. (1994), 'Liability for Information and Opinions Causing Pure Economic Loss to Third parties: Comparison of English and German Case Law', in Markesinis, B. (ed.) *The Gradual Convergence* (Oxford: Oxford University Press)

Wagemans, T., *An Introduction to the Labelling of Websites*, background paper for DG Information Society Conference 'Quality Labels for Websites Alternative Approaches to Content Rating', 27 February 2003, Luxembourg. Available at <http://www.europa.eu.int/information_society/activities/sip/docs/pdf/reports/qual_lab_bkgd.pdf>

Wakefield, R.L. (2001), *A Determination of the Antecedents of Online Trust and an Evaluation of Current Web Assurance Seals*, Doctoral dissertation, University of Mississippi, MS

Wakefield, R.L. & Whitten, D., 'Examining User Perceptions of Third-Party Organization Credibility and Trust in an E-Retailer', *Journal of Organizational and End User Computing* 18 (2) (2006)

Walton, C. et al. (2006), *Charlesworth & Percy on Negligence* (London: Sweet & Maxwell)

Weatherill, S. & Bernitz, U. (2007), *The Regulation of Unfair Commercial Practices under EC Directive 2005/29. New Rules and New Techniques* (Oxford and Portland, Oregon: Hart Publishing)

Winier, H.B., 'Common Law Liability of the Certified Public Accountant for Negligent Misrepresentation', *San Diego Law Review* 233 (1983)

Witting, C. (2004), *Liability for Negligent Misstatements* (New York: Oxford University Press)

Wolfson, J., 'Electronic Mass Information Providers and Section 522 of the Restatement (Second) of Torts: The First Amendment Casts a Long Shadow', *Rutgers Law Journal* 29 (1997)

Zweigert K. & Kötz, H. (1996), *Einführung in die Rechtsvergleichung* (Tübingen: Mohr)

Zweigert, K. & Kötz, H. (1998), *An Introduction to Comparative Law* (New York: Oxford University Press)

INFORMATION TECHNOLOGY & LAW SERIES

1. E-Government and its Implications for Administrative Law – Regulatory Initiatives in France, Germany, Norway and the United States (The Hague: T·M·C·Asser press, 2002)
   Editor: J.E.J. Prins / ISBN 978-90-6704-141-6
2. Digital Anonymity and the Law – Tensions and Dimensions (The Hague: T·M·C·Asser press, 2003)
   Editors: C. Nicoll, J.E.J. Prins and M.J.M. van Dellen / ISBN 978-90-6704-156-0
3. Protecting the Virtual Commons – Self-Organizing Open Source and Free Software Communities and Innovative Intellectual Property Regimes (The Hague: T·M·C·Asser press, 2003)
   Authors: R. van Wendel de Joode, J.A. de Bruijn and M.J.G. van Eeten / ISBN 978-90-6704-159-1
4. IT Support and the Judiciary – Australia, Singapore, Venezuela, Norway, The Netherlands and Italy (The Hague: T·M·C·Asser press, 2004)
   Editors: A. Oskamp, A.R. Lodder and M. Apistola / ISBN 978-90-6704-168-3
5. Electronic Signatures – Authentication Technology from a Legal Perspective (The Hague: T·M·C·Asser press, 2004)
   Author: M.H.M. Schellekens / ISBN 978-90-6704-174-4
6. Virtual Arguments – On the Design of Argument Assistants for Lawyers and Other Arguers (The Hague: T·M·C·Asser press, 2004)
   Author: B. Verheij / ISBN 978-90-6704-190-4
7. Reasonable Expectations of Privacy? – Eleven Country Reports on Camera Surveillance and Workplace Privacy (The Hague: T·M·C·Asser press, 2005)
   Editors: S. Nouwt, B.R. de Vries and J.E.J. Prins / ISBN 978-90-6704-198-0
8. Unravelling the Myth Around Open Source Licences – An Analysis from a Dutch and European Law Perspective (The Hague: T·M·C·Asser press, 2006)
   Authors: L. Guibault and O. van Daalen / ISBN 978-90-6704-214-7
9. Starting Points for ICT Regulation – Deconstructing Prevalent Policy One-Liners (The Hague: T·M·C·Asser press, 2006)
   Editors: B-J. Koops, M. Lips, J.E.J. Prins and M. Schellekens / ISBN 978-90-6704-216-1
10. Regulating Spam – A European Perspective after the Adoption of the E-Privacy Directive (The Hague: T·M·C·Asser press, 2006)
    Author: L.F. Asscher / ISBN 978-90-6704-220-8
11. Cybercrime and Jurisdiction – A Global Survey (The Hague: T·M·C·Asser press, 2006)
    Editors: B-J. Koops and Susan W. Brenner / ISBN 978-90-6704-221-5
12. Coding Regulation – Essays on the Normative Role of Information Technology (The Hague: T·M·C·Asser press, 2006)
    Editors: E.J. Dommering and L.F. Asscher / ISBN 978-90-6704-229-1

13. Customary Law of the Internet – In the Search for a Supranational Cyberspace Law (The Hague: T·M·C·Asser press, 2007)
    Author: P.P. Polánski / ISBN 978-90-6704-230-7
14. Fighting the War of File Sharing (The Hague: T·M·C·Asser press, 2007)
    Authors: A.H.J. Schmidt, W. Dolfsma and W. Keuvelaar / ISBN 978-90-6704-238-3
15. Constitutional Rights and New Technologies – A Comparative Study (The Hague: T·M·C·Asser press, 2008)
    Editors: R.E. Leenes, B.J. Koops and P. De Hert / ISBN 978-90-6704-246-8
16. Self-Regulation in Cyberspace (The Hague: T·M·C·Asser press, 2008)
    Author: J.P. Mifsud Bonnici / ISBN 978-90-6704-267-3
17. Trustmarks in E-Commerce – The Value of Web Seals and the Liability of their Providers (The Hague: T·M·C·Asser press, 2009)
    Author: P. Balboni / ISBN 978-90-6704-296-3

Lightning Source UK Ltd.
Milton Keynes UK
UKOW04n1459280616

277266UK00001B/14/P